TABLE OF CO

CW01508261

Introduction by Brian Becker	VII
1. Historical Background	1
2. From the May Fourth Movement to Liberation	21
3. Building a New China and the Struggle Between Two Lines: 1949-1976	37
4. Reform and Opening Under Deng Xiaoping	61
5. China in the Twenty-First Century	75
Reflection: The Contradictions of Liberation by Eugene Puryear	93
Glossary of Names	129
Endnotes	153
Index	164

CHINA'S REVOLUTION AND THE QUEST FOR A SOCIALIST FUTURE

1804 books

First published in April 2023 by
1804 Books, New York, NY

1804Books.com

ISBN: 978-1-7368500-8-4
Library of Congress Control Number: 2023936203

Cover: Hannah Craig

CHINA'S REVOLUTION AND THE QUEST FOR A SOCIALIST FUTURE

Written By
Ken Hammond

Introduction By
Brian Becker

Reflection By
Eugene Puryear

INTRODUCTION

BY BRIAN BECKER

In the winter of 2008 the Party for Socialism and Liberation (PSL) published the book *China: Revolution and Counterrevolution*. That book outlined a distinctive position and orientation within the US Marxist Left toward the People's Republic of China (PRC). The book was published thirty years after China publicly initiated a radical reorientation of its economic system and foreign policy, frequently referred to as the Opening Up.

China is huge, complex, and dynamic. The 2008 book, however, had a very specific and limited objective. Its audience was relatively limited, too. It was essentially a polemic against those in the Marxist movement—including those identifying as Maoists—who decided, 1) capitalism was largely restored in China, 2) the Chinese bourgeoisie was essentially controlling the state and, 3) the Communist Party of China (CPC) itself had been captured by the bourgeoisie and its continued use of Marxist jargon was nothing more than a cover for a counterrevolution that seized power from the working class and poor peasants following Mao Zedong's death.

It wasn't just the Maoist Left that described China as a capitalist country during this period. Social democracy and social democratic-oriented Marxists in academia also denounced China as a new variant of capitalism, and they condemned the CPC as oppressors of the working class. Importantly, many from these currents denounced Mao and China before the 1978 economic reforms and before the Opening Up to foreign capitalist corporations. Simply put, social democrats, or many of their leading lights, were hostile to the Chinese Revolution both before Mao's death and after his successors changed the economic model.

The idea that China abandoned the socialist road and embraced capitalism was also widely promoted by US political establishment think tanks, academics, the US media, and capitalist politicians. Conventional wisdom asserted it was only a matter of time until the final denouement of Chinese communism. In the Soviet Union, socialism was toppled in 1991 through the dismemberment of the Union of Soviet Socialist Republics (USSR) and the overthrow of the communist-led governments in Eastern Europe, and the CPC was also dissolving socialism and embracing capitalism. The only caveat, according to this popular and dominant narrative, was that the CPC was retaining state power while the communist parties in the USSR-led socialist camp lost their power. Either way socialism was gone or disappearing. Capitalism was on the rise once again. History was ending—capitalism was indeed the crowning achievement of the species. Or so the story went.

The PSL took a different stance. After a two-year-long study and discussion, we published an internal document in 2007, "For the Defense of China Against Counterrevolution, Imperialist Intervention, and Dismemberment." This was a detailed review of the progress made since the 1949 revolution and the contradictions facing contemporary China.

This new PSL book, *China's Revolution and the Quest for a Socialist Future*, has a different thrust. Its core position is the same as the 2008 publication, but its function is to provide a historical survey of the Chinese Revolution. Its audience is wider. It retains an inherent polemical edge because the core political premise is reaffirmed. In that sense, it's both a survey and a specific interpretation of various stages of the revolution and the important debates that took place within the CPC leadership both before and after the 1949 victory.

Since the writing of the 2008 book the political landscape has undergone a profound change, confirming the validity of the PSL's 2008 thesis. The new book has the same framework but provides a crucial update that considers the earth-shaking political changes between 2011 and 2022. But it isn't just an update. It's written for a wider audience. It provides an accessible survey of the history of the Chinese Revolution from a Marxist perspective.

China Turns Left . . . Again!

In 2011 the Obama administration announced the so-called Pivot to Asia, and in 2013 Xi Jinping became the general secretary of the

CPC. The Asia pivot was not widely understood at first. But it inaugurated a decade of growing belligerence and hostility by US imperialism toward China. By 2018 the Pentagon announced a new military doctrine making confrontation and preparation for war with China the center of US military strategy. The earlier period of cooperation gave way to what is now called a New Cold War.

During this same decade the CPC moved politically to the left. The Xi leadership reaffirmed the socialist and communist goals of the party while it carried out an anti-corruption campaign. This constituted a purge of party and government officials, many of whom maintained the rhetoric of socialism while abandoning its key tenets. The Xi leadership also acted forcefully to rein in the growing economic and political power of the capitalist class that emerged inside China since the country's sweeping economic reforms were introduced, following Mao's death in the mid-1970s.

China's ability to shift to the left internally is an important indicator that the sociological or class character of the PRC is the same as it was in 1949 when the Communist Party came to power after a twenty-seven-year-long civil war. Is it a pure class dictatorship of the working class and poor peasants? No. Was it ever? No. But, is China sociologically different from the capitalist countries? Indeed it is, and the Chinese revolutionary project lives on. Its continued existence is crucially important to the peoples of China and the global struggle against imperialism. This was the cornerstone of PSL's 2008 thesis. But it was presented before the leftward shifts under Xi's leadership and before the reorientation of US military and foreign policy following the Asia pivot provided clear proof.

What is the Best Way to Defend China?

The PSL's political position on China has been distinctive since its inception. The party offered both a strong defense of the Chinese Revolution's continued march forward, and a well-rounded recognition and explanation of a primary contradiction with the economic model adopted in the late 1970s. The Opening Up reforms presented a huge boon to China's economic growth, but also created an extreme danger of regression, or capitalist restoration, and possible counterrevolution.

China's harshest critics on the left adopted the line that there was nothing to defend in the country. In contrast, many of China's leftist

China today (Map: Tina Duong)

supporters became cheerleaders for nearly every policy pursued by the CPC. The PSL took a different stance from both.

The PSL argued that the militant defense of China should be premised on an independent and open-eyed assessment of a complex, contradictory process in the most populous country of the world as it sought to achieve, in a short historical period, the daunting task of overcoming the legacy of enforced underdevelopment. We maintained that an idealistic or uncritical approach toward China was not a strong defense. Idealism, once it crashes on the hard rock of reality, usually morphs into cynicism for the idealist. A stable, political defense of China must be rooted in a materialist appreciation of the contradictions of building socialism in a society that was not among the advanced capitalist countries, but rather had been plundered by them.

The PSL explained the function of allowing capitalist market relations to flourish in China was to acquire technology and carry out economic development. But we reject the concept of the capitalist market as a form of socialism. Socialism ultimately means the abolition of private property and the production of goods and services to meet individual and societal needs. The market is the process by which the production of goods and services present themselves as commodities to be bought and sold. A capitalist market is where surplus value—profit—is realized or made for private owners of the means of production. A commodity market, if allowed to flourish, grows spontaneously, and with it grows class divisions. The gamble made by the CPC in 1978 was that the state would be strong enough to retain decisive, supervisory control over the domestic capitalist market. Likewise, as long as the party could maintain state power, it would navigate the complex and treacherous relations with the economic centers of global capitalism to prevent the country from being reduced to a neocolonial status.

Lenin and Deng—Two Political Lines on Market Reforms

The PSL also took note of how the CPC officially explained the Opening Up reforms. The language used was somewhat vague but has profound political significance. The reforms represented the continuation or perhaps the conclusion of a long-standing two-line struggle inside the summits of the CPC. During the first phase of the Cultural Revolution (1966-69) Mao labeled long-time leaders Liu Shaoqi and Deng Xiaoping as "capitalist roaders." It was unclear exactly what that term meant, but this two-line struggle came to a climax when Mao died in 1976.

Following his death Mao's closest collaborators were arrested. The most well-known included top leaders from the Cultural Revolution, including Jiang Qing, Mao's wife. Within two years the new leadership, under the tutelage of Deng Xiaoping, embarked on the policy of radical economic reform. Many of his followers outside China concluded that since the "capitalist roaders" had come to power, Mao's forecast and warning during the Cultural Revolution had come to pass. The capitalist roaders were in power and they were indeed establishing a private capitalist sector. Foreign capitalist corporations were given entry to set up shop, hire Chinese labor at low wages, and expatriate profits derived from the exploitation of Chinese labor. Many leftists outside the country concluded China was socialist in words only, that a counterrevolution had returned capitalism to power, and the new party leaders employed the language of "socialism with Chinese characteristics" as a way of covering their tracks.

PSL rejected that notion. There was no counterrevolution. The new leadership was employing a particular economic model to overcome the legacy of underdevelopment and poverty and modernize the country. China was opening the door to major transnational capitalist corporations that would invest in China, and hire and exploit low-wage labor. In return the country would get access to modern technologies, capital inputs, training, and employment opportunities for hundreds of millions of people. Similarly, the government would modify existing rules and laws to allow the development of a domestic capitalist market and thus allow the creation of a new class of Chinese entrepreneurs and capitalists. But communists would retain control of the state and government policy, overseeing the process with the ultimate power of steering investment to fulfill government economic plans. That which was privatized could be renationalized and transformed into public property. The coercive power of the communist-led state would be employed to maintain control over domestic capitalists.

V. I. Lenin and the Bolsheviks adopted similar capitalist reforms in 1920 to overcome staggering poverty and economic shortages following the Russian Revolution and a three-year-long civil war that devastated the country and its economy. The Bolsheviks invited western capitalist corporations to invest in socialist Russia, employ Russian labor, and expatriate profits back to their investors. Lenin described the New Economic Policy (NEP) as a step backward, a retreat from socialism, but necessary to kick start an economy where people were dying of hunger.

In contrast, the CPC described the post-Mao economic reforms not as a necessary retreat from socialism, but as "socialism with Chinese characteristics." A mixed economy, part socialist and part capitalist, under the party's direction, was not an emergency measure such as Lenin's NEP but rather a preferred long-time economic model.

In the decades following the Opening Up reforms of 1978, the Chinese economy grew at dizzying levels. Urbanization of a country that had been predominantly rural and peasant before 1978 was carried out at a tempo unmatched in human history. While the economic achievements were unmistakable and recognized universally, characteristic features of capitalism also grew. Income inequality between rich and poor, workers and capitalists, soared. Private ownership of companies allowed a tiny part of the population who did not really work to become rich from the profits derived from the labor of millions of people working long hours each week for wages.

Transitions to Socialist Society:
China and USSR—Two Models of Economic Growth

China's great economic leap under Deng Xiaoping's leadership in the 1980s and 1990s used different methods from those employed by the Communist Party of the Soviet Union which had a similar dynamic period of industrialization and urbanization in the 1930s. China in the later decades, and the USSR in the 1930s, experienced growth rates far outstripping those of the major capitalist economies.

The Soviet Union's stupendous industrial growth rates in the 1930s were motivated with the same urgency as that of the CPC leadership. It wasn't just a matter of overcoming poverty and underdevelopment. Both the USSR and China were confronted with a hostile world environment dominated by advanced capitalist governments' efforts to destroy the socialist projects in their respective countries.

Joseph Stalin famously stated about the USSR: "We are fifty or a hundred years behind the advanced countries. We must make up this gap in ten years. Either we do it or they will crush us." He said this ten years before the Nazi invasion of the USSR in 1941.

The Soviet Union's stupendous growth in the 1930s was based on a different model than that pursued by China since the late 1970s. The NEP ended in 1928, and the leadership embarked on a new economic model—based on almost complete state ownership of the means of production and the rapid, forced collectivization of the nonsocialist countryside.

The USSR's economic model of the 1930s precluded the capitalist class. It also excluded foreign capitalist corporations and banks. All the core means of production and finance were state-owned. Central economic planners dictated all elements of production and distribution in the industrial sector and much of the agricultural sector.

In contrast, China's Opening Up reforms included vast privatization. They allowed for the emergence of a new, indigenous capitalist class, but one whose political and economic power would be constrained by the Communist Party-led state and government.

Is China a Socialist Country?

A question that is frequently asked is whether China is a socialist country. Although the economic model adopted by the USSR in 1928-29 was vastly different from the one adopted by China after the Opening Up, the same question was often posed in theoretical and political disputes within the Left.

Marxism requires the precise use of language in addressing complex political and sociological questions. In 1917 in Russia and 1949 in

China, existing capitalist state power was smashed by victorious revolutions and a new state power was created. In both instances the ruling political power was assumed by Marxist-led communist parties professing socialist goals and aspirations. Did that mean they were socialist countries? From PSL's viewpoint the victorious revolutions were a qualitative step forward for both countries on the path to socialism. But the final victory of socialism—eradicating class divisions and other manifestations of inequality rooted in the very nature of existing class society—would be the consequence of a longer historical *process*. The PRC defines its development as still in an early stage of this process. There is nothing in this process, in this transitional era, that offers a rock-solid guarantee of the ultimate achievement of a truly *socialist* society. The outcome will be determined by the struggle of class forces inside China and in the arena of global politics.

In capitalist countries competing ruling-class parties take turns running the affairs of the capitalist government. In the US Democrats and Republicans take turns at the helm. In recent elections in capitalist Germany the center-right Christian Democrats lost power to the center-left Social Democrats. The transfer of authority over the capitalist government between competing capitalist parties possesses no intrinsic threat to the status quo. The political superstructure in capitalist countries plays an important role, but capitalist profit-making and exploitation don't require state oversight and direction.

In the transitional era following a socialist revolution, however, when socialism remains an aspiration and will only finally be achieved over a longer historical period, the role of the superstructure is decisive. *Thus, the retention of state power by the Communist Party remains the indispensable foundation for maintaining the superstructure. In fully developed socialism—what Marxists describe as the emergence of a classless society—political parties representing differing class forces exit the historical stage.* This doesn't happen where class struggles still rage, or in a world where socialism takes the form of isolated nations or clusters of nations fighting to fend off counterrevolutionary threats from powerful imperialist powers.

The Pivot to Asia announced by President Obama in 2011 was a signal that US imperialism is determined to crush socialism in China. Since then US foreign and military policy has been dramatically reshaped to prepare for an all-out confrontation with China. This is already a dominant feature of contemporary global politics. All the more reason for Marxists living in the US to understand China comprehensively, to have a dialectical appreciation of its achievements and challenges, and to stand in defense of the Chinese Revolution which continues to shape and shake the foundations of the modern world.

CHAPTER ONE

HISTORICAL BACKGROUND

The Old Order

In 1793 King George III of Great Britain sent a diplomatic mission to China led by Lord George Macartney. Macartney carried a letter from the king to the ruler of the Qing dynasty, known as the Qianlong emperor, requesting that China grant broad access to its markets to British traders, and allow them to establish permanent diplomatic offices in Beijing. The Qianlong emperor received Macartney in an audience at the Qing Summer Palace in Chengde, northeast of the capital, and gave the British emissary a letter in reply. In it, he politely but firmly declined the British requests, noted that China had everything it needed, and the British had nothing to offer in trade. Further, the existing system of international commerce, handled through the single port of Guangzhou in the far south of the empire, was perfectly fine and would continue to allow the British to purchase whatever Chinese goods they desired. China was the richest, most populous, and most economically advanced country in the world, and the emperor was confident it would remain that way.[1]

Qianlong could not have been more wrong. At the end of the eighteenth century, China was still the wealthiest land on the planet, with 20 percent of the world's population generating over 30 percent of global economic production. But that was about to change dramatically for a combination of domestic and international reasons.[2] China faced rising contradictions in its domestic economy and political system, and the Industrial Revolution in Britain was soon to provide the West with both a tremendous new capacity for large-scale and

Lord Macartney in audience with the Qianlong emperor, 1793 (Drawing: William Alexander)

low-cost production of goods and modern military technologies that allowed Europeans to project force in unprecedented ways and impose imperialist domination on much of the world.

China's imperial system had a history of more than two thousand years, with Chinese civilization extending back another two millennia before that.[3] Beginning in the tenth century, China had what's been called a commercial revolution: the old system of an agrarian economy of mostly local self-sufficiency with an aristocratic elite gave way to an empire-wide network of markets and regional centers of production. Textiles were the main product of eastern China; ceramics came from the great kiln city of Jingdezhen; paper and printing were concentrated in Fujian province; while Sichuan, Hunan, and the North China Plain were great grain producers. Long-distance trade across the empire and into the global economy flourished. Land ownership remained the base of the imperial ruling class, the literati, or the gentry. But a new class of merchants, largely based in the burgeoning cities along the coast and up the Yangzi River valley, arose. Agriculture became increasingly commercialized as production for the market supplanted merely local modes of consumption.[4]

By the twelfth century, China had its early form of capitalism, with commodity production, a highly monetized economy, significant employment of wage labor, an extensive and complex network of markets linked by well-established trade routes, and sophisticated forms of credit and finance serving the needs of both merchants and capitalist manufacturers. Its economy was most highly developed in the Jiangnan region, the Yangzi River delta between the modern cities of Nanjing and Shanghai. Still, the commercial revolution's effects were felt throughout the empire.[5]

As is often the case in developing economies, the textile industry was the leading sector in this evolving system. Urban-based investors in cities like Suzhou or Songjiang used a "putting-out system" to organize the growing, spinning, and weaving of silk and cotton cloths. Most of this activity took place at the household level, but some larger production centers also began to emerge. Workers in Suzhou began to contend with employers, and strikes broke out from time to time, indicating the early development of class struggle.[6]

The development of a capitalist sector in China did not follow the same patterns as the later emergence of capitalism in Europe. Over the course of the period from the tenth through the eighteenth centuries, China's economy went through major upheavals and periods of contraction and devolution: during the Mongol conquest in the thirteenth century, the dynastic transition from the Mongol Yuan dynasty to the Ming in the late fourteenth century, and the shift from the Ming to the Qing dynasty in the seventeenth century. At these times, there was massive destruction of property and loss of life. But each time the dynamics of the market reasserted themselves once order was restored, and the Ming dynasty saw a tremendous expansion of the commercial sector in the sixteenth and early seventeenth centuries. The final flourishing of China's early capitalism extended through the eighteenth century to the era of the Qianlong emperor.

The history of China's political economy diverged from that of Europe in other ways, though there were some important parallels. Perhaps most significant in terms both of the nature of its early modern history and its implications for China today, was the relationship between commercial or manufacturing elites and the older land-owning elite that had long dominated access to political power through its exclusive control of the imperial examination system—the primary means of recruitment for officials in the bureaucratic administration. As China's economy became more commercialized, with both town-centered markets and centers of manufacturing production, the agricultural sector also became increasingly drawn into production for the market, commodity production. Landowning families had to adapt to the demands and functions of the market, but they also began to accumulate greater wealth and sought ways to invest that money. Merchants and manufacturers, themselves accumulating significant wealth, sought social prestige and legitimation by emulating the cultural practices of the literati elite. Literati invested discreetly in commercial enterprises, while successful capitalists invested in land to elevate their social status and respectability. These economic strategies resulted in more of a convergence of interests between landowning and commercial elites than an antagonistic relationship, in contrast to the hostility and conflict between bourgeois and aristocratic elites in European history.[7]

This convergence of class interests is not unlike that described by Marx in "Review of Guizot's Book on the English Revolution" in the February 1850 *Neue Rheinische Zeitung Revue*. Discussing what he called the conservative nature of the English Revolution of 1688, Marx argued it was shaped by an alliance between the English commercial and industrial bourgeoisie and the increasingly commercialized large landed-estate owners. He wrote:

> This class of large landowners allied with the bourgeoisie . . . was not, as were the French feudal landowners of 1789, in conflict with the vital interests of the bourgeoisie, but rather in complete harmony with them. Their estates were indeed not feudal but bourgeois property. On the one hand, they provided the industrial bourgeoisie with the population necessary to operate the manufacturing system, and on the other hand, they were in a position to raise agricultural development to the level corresponding to that of industry and commerce. Hence their common interests with the bourgeoisie: hence their alliance.[8]

Both sides of this ruling-class collaboration of course remained dedicated to the extraction of surplus from the labor of workers, whether on farms, in homes or workshops, or the marketplace, and to the accumulation of wealth in their own hands. Class struggle between landed and commercial/industrial elites may have been minimal. Still, resistance to exploitation by the rich and powerful was a regular feature of imperial history, with strikes and other labor actions in the towns, and rebellions in the countryside recurring across the long centuries of dynastic rule.

The imperial system survived, despite the disruptions and upheavals of dynastic transitions, because this convergence of elite interests was also articulated in the political order. While individuals registered as merchants were officially prohibited from taking imperial examinations and serving in government positions, the state nonetheless promoted and protected the operations of the commercial economy. Even Confucian scholars, who had traditionally enunciated an aversion to wealth generated through trade and the pursuit of profit, from the tenth century on, came to new understandings and interpretations of Confucian ideology, which legitimized and even valorized the contributions of merchants and manufacturers to the needs of society.[9] Imperial officials viewed the actions of these economic agents as contributing to the flourishing of the empire. The state sought to constrain excessive displays of wealth that could provoke unrest, but encouraged productive investment and wealth accumulation as promoting a stable and prosperous order.[10] In some ways this orientation of the historical political economy foreshadows the hybrid complexity of contemporary China's economic system.[11]

Domestic Crisis and Imperialist Assault

By the late eighteenth century, China faced the limitations and contradictions of its economic and political situation. As the nineteenth century got underway, new challenges arose from the outside world as well. Within the empire, the development of productive capacity in agriculture and manufacturing reached a high level, but combined with population growth, the potential for further accumulation was declining dramatically. Pressure on resources eroded the material standard of living for tens of millions of Chinese. The late eighteenth century saw the first popular uprisings in more than a hundred years, and this trend continued and grew after the turn of the century. As productivity stagnated, government revenues declined, and the administrative capacity of the state eroded. Lack of maintenance of basic infrastructure such as the Grand Canal, which facilitated trans-

British ships attack Chinese naval defenses.

port of grain from the country's south to the north to feed the large population around the imperial capital of Beijing, caused economic distress for many parts of the empire. The 1799 death of the Qianlong emperor was followed by a series of mediocre emperors who proved unable to respond effectively to the deepening crisis.[12]

As the old order faced rising contradictions, there were changes on the other side of the world that would lead to a new era of global power relations, and the assault of Western imperialism on China and the rest of the world. Britain's Industrial Revolution at the end of the eighteenth and into the nineteenth century fundamentally transformed not only the productive capacity of British and European manufacturing, but also gave rise to new developments in communication and transportation, and new military technologies. These allowed first Britain and then other Western countries to project power in unprecedented ways, conquering lands and peoples in Asia and later Africa, and creating the colonialist and imperialist order dominating the planet for much of the nineteenth and twentieth centuries.

The rise of British imperialism in China was closely intertwined with the conquest of India by the English East India Company (EIC), a mercantilist-chartered corporation that had operated in the Indian Ocean arena since the early seventeenth century.[13] As the Mughal empire—which had ruled the subcontinent for more than two hundred years—faltered in the middle 1700s, the EIC maneuvered into positions of strength by intervening in internal conflicts, especially in the Bengal region. India had the world's most advanced cotton textile industry, and Bengal was a major production center. Cotton grown locally was worked up in a highly sophisticated system of spinning, weaving, dyeing, and printing, then marketed worldwide, from Japan and Southeast Asia to Europe, Africa, and the New World.

In 1757 the EIC took part in the Battle of Plassey and gained control of Bengal. At first this led to the simple looting of the province, as the EIC extracted "gifts" and "tribute" from local elites. But as the Industrial Revolution got underway, English capitalists developed their textile industry. So, the EIC systematically destroyed the Indian cotton economy, drove farmers out of cotton production, and shut down spinning and weaving operations. British industry bought cotton grown on slave plantations in the southern United States. As India was eliminated as a competitor, the mills of the English Midlands flourished.

Destruction of the Indian textile industry was an imperialist crime in its own right. But it also set in motion an even greater criminal enterprise. As the EIC forced farmers to stop growing cotton, they encouraged them to cultivate opium poppies. Soon, far more opium was being produced than could be consumed for legitimate medicinal purposes, so the EIC looked for new markets to distribute the drug. They soon learned of potential demand among the Chinese and shipped larger and larger volumes.

Opium radically altered the economic relationship between Britain and China. This was when the Macartney mission failed to open China to British trade, when the Qianlong emperor pointed out that the British had no goods the Chinese wanted. The rise of the opium trade led to a reversal in the balance of payments, with silver flowing out of China to pay for the opium being smuggled into the country.[14]

Opium was illegal in China, so the British couldn't bring it in via the normal trade system in Guangzhou. Instead, they colluded with Chinese merchants to establish an elaborate network of remote bays and inlets in the Pearl River Delta, where British ships offloaded opium before going on to Guangzhou for their legal trade. Huge amounts of opium flowed into China, draining silver from the domestic economy and causing major problems of crime and social disruption. Shipments grew steadily through the 1820s and 1830s.

The Qing government sought ways to control the situation, and in 1837 appointed an official, Lin Zexu, commissioner for opium control at Guangzhou. Lin tried to suppress the trade, but the British resisted and called upon Parliament to come to their defense.[15]

In London, Parliament debated the China situation. Of course, they did not talk about the interests of drug dealers as the issue, but argued it was a matter of free trade, that a British merchant should be able to go anywhere and trade anything with anyone. The Qing government was blocking free trade, and needed to be taught a lesson. Parliament voted to launch the Opium War. From 1839-42, the Royal Navy cruised up and down the south China coast, shelling port cities, and killing thousands of innocent Chinese. The military superiority of British modern industrial weaponry quickly destroyed effective resistance by Chinese forces.

Qing negotiators tried to pacify the British, but the foreigners wouldn't settle for anything short of complete surrender. In 1842, the Qing government was forced to sign the Treaty of Nanjing, the first of the unequal treaties which would build the system of imperialist domination over China.

The treaty made China open five ports on the southeast coast to British traders and missionaries. Over the following decades many more "treaty ports" would be opened across the country. Britain set the terms of tariffs and duties. The treaty also established the principle of extraterritoriality, meaning that British citizens who committed crimes in China could not be prosecuted under Chinese law, but must be returned to British control for any legal procedures. In effect, British citizens were immune to local jurisdiction. A final clause of the treaty provided there would be no "most favored nation," meaning that if any other country signed a treaty with China in the future in which China granted them a concession or privilege not included in the Treaty of Nanjing, those provisions would automatically apply to Britain as well. The Treaty of Nanjing, supposedly an agreement between equal diplomatic partners, was instead a document of domination and subordination, legally enshrining China's military and political humiliation.[16]

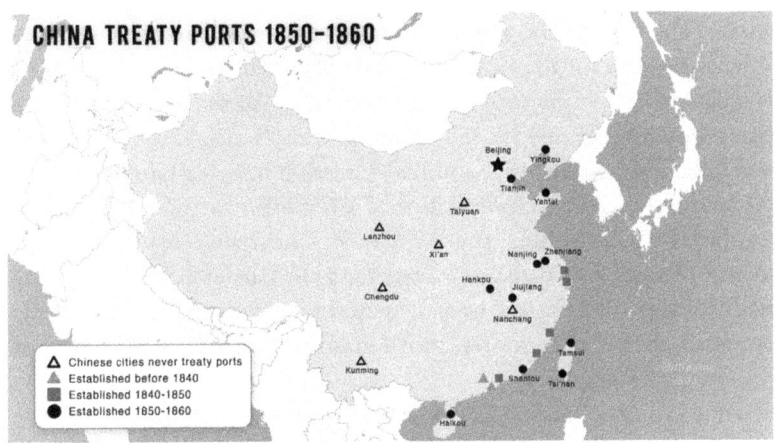

Treaty ports spread across China (Map: Tina Duong)

The Century of Humiliation

From 1842 until 1949, China endured more than a hundred years of domination and exploitation by Western imperialism. This period saw the decline and near collapse of China's domestic economy as mass-produced, low-cost products from Western factories poured into the country, undercut the prices of goods from China's traditional manufacturing sector, and displaced millions of workers. As time went by, Western capitalists built their own factories, mostly in or near Shanghai, giving rise to the emergence of a modern industrial working class. But the economy overall remained overwhelmingly agrarian. As China became more integrated into and subordinated to the new global economic order, the mass of Chinese farmers fell on hard times, and the standard of living for tens of millions declined dramatically.[17]

The impact of Western imperialism on China very quickly generated resistance and rebellion. By the 1850s massive uprisings broke out in many parts of the country.

The greatest uprising was the Taiping Rebellion, lasting from 1853 to 1864. It drew more than twenty million people into the struggle and at its peak affected nearly a fifth of China's territory. The Taiping movement grew out of economic hardship among the minority Hakka people in southern China, many of whom lost their employment as porters and haulers of goods to the markets of Guangzhou in the wake of the Treaty of Nanjing and the opening of new coastal ports farther

9

north. The movement was led by a man named Hong Xiuquan, a failed candidate for the Confucian civil examinations. After a nervous breakdown caused by repeated frustrations with the exams, he had a vision in which he saw the Christian God and Jesus, who told him he was Jesus' younger brother and had a mission to bring Christianity to the masses of China. Hong launched his movement with a dream of emulating the communalism of the early Christians, creating an ideal society of shared wealth and overturning the patriarchal bonds of the traditional family. Beginning as a small rural commune in Guangdong province, the movement grew into a massive challenge to the power of the Qing dynasty.[18]

In 1853, the Taiping began a march to the north to overthrow the Manchu rulers and establish the *Taiping Tianguo*, the Heavenly Kingdom of Great Peace. They fought with government forces, but also drew greater and greater numbers of supporters from the poor communities they passed through. They set up their kingdom with its capital at Nanjing, a former imperial capital on the Yangzi River, not far from Shanghai. For the next decade Nanjing was their central base, with Taiping-controlled territory stretching across much of southern China. Unfortunately, contradictions emerged within the movement, as leaders indulged in a more luxurious lifestyle, while the mass of Taiping followers continued trying to create a more egalitarian society. This undermined the cohesion of the Heavenly Kingdom and weakened the ability of the Taiping to fight the Qing armies. By 1864, Nanjing fell to government forces.

The Taiping Rebellion was the largest of the anti-Qing uprisings, but by no means the only one. Popular rebellions broke out in different parts of China, sometimes among non-Chinese communities such as Muslims in Yunnan province, or in especially poor farming areas like Shandong province. The Qing government managed to bring all these challenges to its power under control, but at the cost of a seriously weakened empire. Western imperialists took advantage of these crises by providing some minimal military support to the dynasty, but also used the threat to Manchu rule as a way to squeeze further concessions from the government. In the midst of the Taiping Rebellion, the British and French fought a Second Opium War in 1859-60 to force the Qing to accelerate the pace of opening new ports and allowing ever greater penetration of China by Western traders and missionaries. A

Ruins of the Summer Palace, looted and burned in 1861.

joint military force occupied the capital and looted and burned the imperial Summer Palace just northwest of the city. Many of the treasures carried away by these imperialist vandals remain in museums or private collections in Britain and France to this day. The ruins of the old Summer Palace are today a public park in Beijing commemorating this low point in China's modern history.[19]

While the Qing dynasty fought against rebellions across the empire and tried to resist further demands of the Western imperialists, a few progressive elements emerged among Chinese officials. These were mostly provincial-level leaders, often with ties to the business community trying to develop a modern domestic response to the industrial capitalism of the British and other Western powers. By the 1870s and 1880s, a Self-Strengthening Movement on the part of a handful of Chinese administrators sought to create a Chinese industrial and military sector that would allow China to stand up to Western imperial power. They founded an arsenal and a naval shipyard. They set up schools to train translators in foreign languages so China could gain direct access to Western scientific and intellectual knowledge, and sent delegations to Europe and America to study how businesses, factories, and schools were run.

These efforts faced resistance from the reactionary majority of Chinese and Manchu officials, who feared any reform or change that might undermine their power and privileges, and did not understand or appreciate the seriousness of the threat presented by the foreign aggressors. In 1885 China fought a short war with France to protect northern Vietnam from French colonial conquest. China lost the war, but did not suffer the same outright military humiliation as in the two Opium Wars. But the efforts of the Self-Strengtheners weren't enough to save the nation from further oppression at Western hands.[20]

While China struggled to find a way to respond to the power of the industrial West, Japan embarked on a radically different path. Japan developed its own early modern variant of capitalism during the Tokugawa era (1603-1867), with a dynamic urban-centered market economy and an increasingly commercialized agricultural sector.[21] When American gunboats forced Japan to open some of its ports to foreigners in 1854, Japanese political leaders quickly decided to embrace Western models and practices to modernize Japan to avoid being colonized and subordinated to Western interests. The Meiji Restoration of 1867-68 put Japan on a fast track to full industrialization and participation in the global capitalist system as one of the new imperialist powers rather than as a victim of the West. This path quickly brought it into conflict with China.[22]

In 1895 Japan fought a war with China over control of Korea. Korea and China had a long relationship of close political and economic links, but Japan's rising imperialist ambitions targeted Korea as an opportunity for colonial expansion. Japan's modernized military defeated the Qing forces on land and at sea, and China was forced to renounce its "special relationship" with Korea. Japanese imperialism continued to focus on China as the twentieth century dawned and plagued East Asian countries down through the middle of the century.[23]

China's defeat by Japan triggered a new phase of political crisis for the Qing. In 1898, a small group of progressive reformers won the support of the Guangxu emperor for a few months, and launched a serious effort to modernize the imperial administration. From June to September, what became known as the Hundred Days Reform, led by figures such as Kang Youwei, Liang Qichao, and Tan Sitong, put into place measures to streamline government operations. It opened up new lines of communication between Confucian bureaucratic officials and ordinary citizens, modernized education, and promoted innovation and creativity in government policies.[24] These reforms, like the Self-Strengthening Movement of the preceding decades, threatened the reactionary majority of imperial officials and the Manchu nobility.

The Guangxu emperor was pro-reform, but was not a strong ruler. Real power lay in the hands of his aunt, the Empress Dowager Cixi, who exercised that power from "behind the screen," out of public sight. On September 21, 1898, the Empress Dowager, in collusion with a general named Yuan Shikai, had the reformist officials arrested and placed the emperor under house arrest. Kang Youwei and Liang Qichao managed to flee, first to Shanghai, then to Japan. But Tan Sitong and seven others were publicly beheaded to make it clear that these kinds of political activities would not be tolerated by the Manchu rulers. This sealed the ultimate fate of the dynasty.

While the Qing political elite struggled over how best to retain its power, the masses of ordinary Chinese continued to suffer the effects of imperialist exploitation and seek ways to resist. One particular source of friction was the activities of Christian missionaries. These agents of cultural imperialism spread across China in the decades after the Treaty of Nanjing with the coerced protection of the Qing state. They fostered fear and resentment among many people, and became a focus for anti-imperialist anger and frustration. In Shandong province in the 1890s, a local martial arts society openly opposed the presence of German missionaries in the villages of the area. A series of attacks on missionaries and their Chinese converts was at first repressed by local Qing officials. By the late 1890s this anti-imperialist movement grew, and the Qing government saw them as potential allies to push back against Western domination. In 1899, after the suppression of the reform movement the previous year, the Boxer Rebellion—named after the martial arts practices of its leaders—became a mass movement, and marched to the imperial capital at Beijing in the spring of 1900 to call for expelling the foreigners from China.[25]

That summer the Boxers then laid siege to the Legation Quarter in Beijing where foreign powers had their embassies and consulates. In response, the imperialist powers, normally contending among themselves individually for maximum advantage, joined together to suppress the Boxers and reestablish control of China. An Eight-Power Expeditionary Force was sent to Beijing, with forces from Britain,

Armies of the Eight-Power Expeditionary Force

France, Germany, the US, Russia, and Japan among them. They broke the siege of the Legations, slaughtering many Boxers and ordinary citizens. They forced yet another humiliating settlement on the Qing government, including the execution of several leading officials, and an indemnity of many millions of ounces of silver, making China pay for the armies that had come to oppress it. This punitive financial burden further deepened the crisis of the Qing state, draining much-needed resources from the country. These obligations were never fully paid, and were only lifted during World War II.

The failure of the Boxer Rebellion and the imperialist occupation of Beijing was the last straw for many, and the fate of the imperial system was sealed. Patriotic Chinese turned away more and more from reformist ideas of modernizing the dynasty to the revolutionary goal of overthrowing it and establishing a modern government. The key figure in this new era was Sun Yatsen, a medical doctor from southern China who had studied in Hawaii. Sun became active in revolutionary circles in the 1890s, and forged a broad alliance of anti-Qing groups called the *Tongmenghui*, which eventually became the Nationalist Party—the Guomindang (GMD), or Chinese Nationalist Party. In the decade after the Boxer Rebellion there were repeated uprisings across the country, especially in the south, and support for the Nationalist revolutionary movement grew steadily. The GMD made special efforts

Sun Yatsen

to organize among the military, and many soldiers and officers secretly joined in the cause.[26]

The Qing dynasty made some minimal gestures toward reform in the first decade of the twentieth century, abolishing the Confucian examinations in 1905, and calling for the gradual introduction of consultative assemblies over the coming years, but these were far too little, far too late. When the Guangxu emperor and Empress Dowager both died in 1908, and a three-year-old boy ascended the throne, the dynasty lost all capacity to save itself.[27]

The end came in October 1911. A military garrison in Hankou, part of the modern city of Wuhan in central China, mutinied when rumors spread that members of the Nationalist movement among the soldiers were about to be arrested. The rebels called upon local literati leaders to head a provisional government and declared their independence from the Qing empire. Military units in provinces in central and southern China soon followed suit, and within a couple of months most of the country had cast off Manchu dynasty rule. Sun, traveling in the US giving speeches and raising money from overseas Chinese communities, received word of the revolution by telegram in Denver. He returned to China at the end of the year to become president of a provisional national government. On February 12, 1912, the last emperor, the little boy Puyi, abdicated the throne. The two-thousand-year-old imperial system was abolished, and China was on the threshold of a new era.

The Nationalists led by Sun Yatsen, hoped to establish a bourgeois republic, with multiparty elections and a constitutional system based on Western parliamentary models. But this was not to be. On February 15, 1912, Sun resigned as president and was replaced by Yuan Shikai, a general in the imperial army who had been commander of the most modernized forces, and who had brokered the abdication agreement with the Qing government. His price for this service was to become president. Over the following months, Yuan subverted the political process, expelled the Nationalists from the newly elected constituent assembly, then had himself made president-for-life. He had the leader of the Nationalist delegates assassinated, and launched a campaign of intimidation against any political actors who opposed him.

While Yuan consolidated his grip on the government, Japan pursued imperialist ambitions, first through the annexation of Korea in 1910 and then with a new focus on China. Japan dislodged China's influence in Korea in the war of 1895, and then slowly expanded its control, making Korea a "protectorate" in 1905, terminated the Korean monarchy and formally annexed the country five years later. Korea would remain a Japanese colony until 1945. Once Japan acquired Korea, China became the next target. When the Western imperialist powers attacked each other in 1914, unleashing World War I, Japan saw an opportunity to advance its interests in China. In 1915, Japan sent a secret diplomatic note to China known as the Twenty-One Demands, seeking special privileges for Japanese businesses and creating a kind of supervision over the government by Japanese "advisors." China initially rejected these demands, but after the most aggressive ones were removed, Yuan's government agreed to grant concessions to the Japanese. Japan's imperialist goals in China continued to develop over the coming years.[28]

Before long, Yuan developed ambitions to found a new dynasty in China, restoring the old imperial system with himself as the new emperor. In April 1916 he attempted to launch his new regime, donning imperial robes and going to the Temple of Heaven to stage the ancient ritual of plowing a furrow to begin the agricultural season. He assembled a small "court" of former Qing officials to legitimize his rule. But his actions were greeted with ridicule or outright rejection by

Cover of *New Youth*, one of the main publications of the New Culture Movement

the people and his imperial hopes soon collapsed. He fled Beijing and died soon after at his home in central China.

With Yuan's death, the political situation in China grew worse. No other figure had the military strength or charisma to dominate the entire country, and China disintegrated into local warlord territories. Military strongmen used forces under their control to carve out local domains for themselves. These were sometimes the same as an old province, sometimes several provinces together, though some were smaller and only comprised a few counties or prefectures. Warlords fought one another to expand their territories, and imposed harsh new taxes on the populations under their control to raise the money for their campaigns. A façade of national government was kept up, with whichever warlord controlled the city of Beijing treated as the head of the Chinese government by the imperialist powers. But in reality there was no government during these years.

The warlord era, 1916-27, plunged hundreds of millions of Chinese peasants into deep distress. With the collapse of the Qing imperial state and the failure to establish a viable bourgeois republic, a combination of military strongmen and local landlords became the effective government in most parts of the country. Their exploitation of the people intensified beyond the norms of traditional society and created the basis for the revolutionary upheaval which was about to break out. In the cities, meanwhile, an emergent Chinese capitalist class, known as compradors for their connections and cooperation with imperialism, colluded with organized-crime groups to dominate business and civic life. Workers in factories, whether owned by foreigners or Chinese, faced severe repression for any efforts to improve their working conditions.

While the material conditions for workers and peasants deteriorated, many educated Chinese were swept up into the movement for a new society. The New Culture Movement arose in the 1910s, growing out of critiques of the old imperial system and its Confucian ideology that began in the late nineteenth century. Scholars and intellectuals like Cai Yuanpei, Chen Duxiu, and Lu Xun attacked the ways the old order held China back and kept it from modernizing to take its proper place in the world. They repudiated everything from the obscurities of the classical written language to the oppression of women in the traditional family, and called for a "new culture" for a "new China." Writers began using vernacular Chinese, the language actually spoken by ordinary people, to make their ideas accessible to the masses. The journal *New Youth* was widely read by young Chinese seeking a path to a better future for themselves and their country.[29]

As these new currents of cultural politics were stirring, the Bolshevik Revolution broke out in 1917, and immediately became a beacon of inspiration for patriotic Chinese. The overthrow of the czarist state, the withdrawal of Russia from the imperialist World War I, the renunciation of secret treaties, a commitment to open and fair international relations, and the revolutionary program of economic and social justice and equality, all aroused Chinese hopes for their country's future. There had been some awareness of Marxist and socialist ideas in China prior to 1917, but the news coming from Russia set off a rapid expansion of interest in revolutionary theory and practice. Reading groups sprang up in cities across the country, and new journals began to circulate translations of the writings of Karl Marx, Friedrich Engels, and V.I. Lenin. In Beijing one study group began at Beijing University, and included a young worker in the university library named Mao Zedong.[30]

By 1919 the situation for China was dire. There was no effective central government, and the depredations of warlords, landlords, and comprador businessmen were bleeding the peasants and workers dry. Western and Japanese imperialists continued to extract massive profits from the labor of farms and factories, and maintained their power and privileges in their concessionary areas in cities across the country, legally immune to Chinese jurisdiction. But the stirrings of the New Culture Movement and the inspiration of the Bolshevik Revolution created new opportunities for the Chinese people to find a path to build a new China. With the end of WWI, the victorious imperialist powers convened a peace conference at Versailles, outside Paris, to punish their defeated foes and establish the terms of their new era of global domination. In China the stage was set for dramatic developments that would set in motion the great revolution of the Chinese people.[31]

FROM THE MAY FOURTH MOVEMENT TO LIBERATION

In the late winter and spring of 1919 leaders of the victorious Allied powers of World War I, Great Britain, France, the United States, and Italy, gathered at the old royal palace complex at Versailles, just outside Paris. They met to settle peace terms to impose on their defeated rivals, the German, Austro-Hungarian, and Ottoman empires, and to decide among themselves Europe's postwar political conformation. While Germany basically retained its prewar boundaries, the Hapsburg and Ottoman empires were dismembered. Their territories were either divided into successor states with ethnic identities, such as Hungary or Czechoslovakia, or into protectorates under the control of either Britain or France, such as Syria or Iraq. The Allies were then the dominant players in the capitalist and imperialist global system. They assumed the rights and powers to make decisions about the ways other peoples would be ruled, not only in Europe and the former empires of the war's losers, but around the world. While the rhetoric of self-determination was cynically deployed during the war to foment unrest and division within the Austrian and Ottoman territories, the victors had no intention of applying such an idea to the colonial possessions they exploited in Asia, Africa, and the Caribbean.[32]

One particular situation to be resolved at Versailles highlighted the solidarity among the imperialist powers. By the end of the nineteenth

century, Japan emerged as a rising power, avoiding Western imperialism's domination by emulating Europe and America's domestic
economic and political development, and their colonial exploitation
of other countries. Having taken Taiwan from China in 1898 and
annexing Korea in 1910, Japan seized the opportunity at the beginning of World War I to grab the German concession at Qingdao, in
Shandong, as well as German outposts in the Pacific. Japan formally
allied with the British, French, and American forces during the war.
So too, had China, which sent tens of thousands of workers to France
to aid the war effort by replacing French men in factories and serving
in supporting roles behind the lines of combat. When the war ended in
November 1918, China assumed Germany's former concession would
be returned to Chinese sovereignty. Instead, the Allies handed it to
Japan, recognizing that country's place as a fellow imperialist power.
China was weak and divided by warlord rivalry, and the government
based in Beijing (then called Beiping) could not stand up to the victors
at Versailles. The Chinese delegates to the Peace Conference agreed
to the treaty's terms, the latest example of China's humiliation at the
hands of the great powers.[33]

News of this betrayal arrived in China by telegram on the morning
of May 4, 1919. As word spread through the capital, students gathered
at the Beijing Normal College campus on the near-west side. By noon
they marched to the Tiananmen—the Gate of Heavenly Peace, the
entry to the old Imperial City—where they held a rally condemning
the Allies' actions, the weakness of their government, and especially
Japanese imperialism's aggressions. Later that afternoon, they marched
to the home of the foreign minister and burned it. Police arrived and
fought with the students, arresting many. This triggered what became
the May Fourth Movement, a protest and boycott directed at Japan
which spread across the country and came to be seen as the beginning
of the revolutionary struggle that led to the establishment of the People's Republic of China in 1949.[34]

The May Fourth Movement grew out of the intellectual and political ferment of the years after the collapse of the old imperial order
in 1911-12. The New Culture Movement helped reform the written
language, making literacy more accessible, and bringing more people
into active participation in public life. The Bolshevik Revolution in
1917 stimulated great interest in Marxist and Leninist thought and

communist organization. The Versailles peace conference led many Chinese to be thoroughly disillusioned with the supposed virtues of Western liberal democracy, exposed as the mere rhetorical façade of imperialist exploitation. The stage was set for the emergence of a new political force that would lead the revolutionary struggle.[35]

Founding of the Communist Party

Through 1919 and 1920 many Chinese formed reading groups to learn about Marxism and the events in Russia. Important figures like Chen Duxiu and Li Dazhao, who'd been involved in the cultural movements of the 1910s, as well as many younger activists such as Mao Zedong, who worked in the Beijing University library at the time, gathered to read and discuss texts by Marx and Engels and the writings of Lenin, newly translated into Chinese. These groups sprang up across the country, in big cities like Beijing and Shanghai and in provincial centers like Changsha and Wuhan. Debates and discussions of capitalism's and imperialism's relevance to China's situation were lively and exciting.

At the same time, Russian revolutionaries understood that the overthrow of capitalism and the destruction of imperialism and colonialism must go hand-in-hand. The Bolsheviks faced serious challenges from counterrevolutionary forces in the civil war. Still, even so, they undertook to provide advice and assistance to radical movements worldwide, especially in Asia. The Communist International, also known as the Third International, was established in Moscow as a center to coordinate and support organizing activities in many countries. Agents were sent abroad to work with local activists and develop new organizations, including forming communist parties, where possible. China became a major focus of these activities, given its size and importance as both a country with a significant economy and a victim of imperialist aggression and humiliation.[36]

China's political situation was complex in the early 1920s. The Guomindang (GMD), the Nationalist Party, led the overthrow of the old imperial system in 1911-12. It was then marginalized by the militarist Yuan Shikai, who made himself president-for-life and in 1916 launched a failed attempt to become a new emperor. After his death in April of that year, the country fragmented into local warlord territories, with the GMD confined to Guangdong province in the far south. Sun Yatsen, the Nationalist leader, espoused ideas including opposition to Western imperialism and a vague kind of socialism called People's Livelihood. The GMD represented the Chinese national bourgeoisie's interests in developing the modern capitalist economy.

In the early 1920s, the Bolsheviks and leaders of the Communist International thought the GMD should be supported as the most viable progressive force. China's industrial sector was a very small component of the overall economy, and the industrial working class was a tiny part of the population, a majority of whom were farmers and agricultural workers. At the same time, the International sought to encourage the formation of a Chinese communist party to be the real leadership of the proletariat, with the long-term prospect of becoming the main revolutionary organization. Sun Yatsen accepted assistance from the International's representative in China, a Dutchman named Henk Sneevliet, known as Maring, who helped reorganize the GMD along Leninist lines of democratic centralism, making it a more effective political operation.[37]

By the summer of 1921, the time was ripe for forming a communist party. July 1 is the official anniversary of the party's establishment, though the first formal meeting of the organization took place a few weeks later. On July 23, twelve representatives from the various local groups that had met over the previous year or two, along with Maring and another International representative, gathered in Shanghai for the founding congress of the Communist Party of China (CPC).

After several days of discussions, the meeting was transferred to a houseboat on a lake in nearby Zhejiang province due to security concerns in Shanghai. Important leaders like Li Dazhao and Chen Duxiu were not present, but Mao Zedong did attend. Mao and another delegate, Dong Biwu, were the only members attending the CPC's founding in Shanghai who were present later at the establishment of the People's Republic of China (PRC) in 1949.[38]

There was a second founding moment for the CPC, held in France in 1922 among Chinese workers and students who had gone to Europe during World War I and stayed into the early 1920s. This group included Zhou Enlai, who went after the war on an academic scholarship, Li Lisan, Cai Hesen, Deng Xiaoping, and others. They became involved with the French Communist Party through work experiences during and after the war. They went on to study in the Soviet Union

Site of the First Congress of the Communist Party of China, July 1921 (Photo: Pyzhou)

before returning to China in the later 1920s to become part of the revolutionary struggle there.[39]

The First United Front

Because of the small size of the working class in China, the Bolsheviks and advisors from the International argued that the new Communist Party should form an alliance with the GMD, to pursue a process of both economic development and building the CPC as the proletariat grew in numbers. This was a contentious point, with many Chinese activists urging a more independent role for their party. In 1923 an agreement was reached for CPC members to join the GMD, but remain under party discipline. Large numbers of communists joined the Nationalist Party. Many rose to important positions within it, including Mao Zedong, who became a director of the Nationalist's peasant bureau, in charge of rural organizing among the farmers of his native Hunan province in central China.[40]

The United Front, as this alliance was known, helped the Nationalists become a more effective and powerful force, and helped create the conditions that enabled the GMD to move beyond its position in

Guangdong province and begin reunifying China under a national government. It also gave many CPC members valuable experience working in different parts of the country, in city factories, and rural villages. As long as Sun Yatsen remained the leader of the GMD, the United Front was a functional policy for both parties. But when Sun died in March 1925, Nationalist leadership passed into new hands, and the political alignment between the two groups frayed.

One of the initiatives that grew out of the GMD-CPC United Front was the establishment of the Whampoa Military Academy, a training center outside Guangzhou. It became an important factor in producing officers who later became leaders of both the Nationalist forces and the Chinese Red Army. Soviet advisors were among the instructors, and several Chinese military leaders went to the Soviet Union for advanced training. Among these was Chiang Kai-shek (Jiang Jieshi), a rising figure within the GMD who later became a close advisor of Sun Yatsen. Chiang was strongly anti-communist, and his time in the Soviet Union did not change his mind. When Sun died, Chiang maneuvered to become the GMD's new leader. His commanding position within the party's military wing gave him a special advantage in the ensuing political struggles.

In 1926 Chiang launched the Northern Expedition, a military campaign to bring the rest of China under Nationalist control. GMD armed forces, including CPC elements, moved north from Guangdong, through Hunan and Hubei provinces, sometimes fighting with local warlords, at other times drawing provincial strongmen into the Nationalist camp through negotiation. By early 1927, Chiang's army had taken Wuhan, on the Yangzi River, and moved east toward Shanghai. As the most industrialized city in the country, Shanghai was the center of the CPC's organization, including large numbers of workers in factories run by foreign imperialists, Western or Japanese, and Chinese capitalists. It was a city ruled mostly by foreign powers in the two major treaty port areas, the International Concession and the French Concession, where Chinese law did not apply. Major organized-crime groups were powerful in Shanghai, further complicating the political landscape.[41]

In April 1927 Nationalist forces reached the outskirts of Shanghai. CPC activists and workers' militias launched an uprising to seize power in the city, welcome the GMD army, and position themselves to play leading roles in the future. But Chiang Kai-shek held his troops outside Shanghai and conspired with crime gangs and imperialist police forces to wage a counterrevolutionary struggle against the workers. Thousands of people were killed, and many thousands more were arrested. Chiang broke with the CPC, and the brutal suppression of the uprising in Shanghai was the perfect opportunity for him to strike against the party and the unions. Once the CPC was suppressed, the GMD army entered the city and established control of the areas outside the foreign concessions. CPC leaders who had not been killed in the fighting were hunted down and summarily executed. The organizational structure of the party was destroyed, and thousands of members were killed. The counterrevolutionary coup led to a split between the GMD and the CPC that set in motion the long struggle for power that lasted for the next twenty-two years.[42]

Seeking a New Path

In the aftermath of the crushing of the CPC in Shanghai, echoed by assaults on party activists elsewhere and the complete rupture of relations between the GMD and the CPC, the Party struggled to reconstitute its leadership, analyze the political situation, and determine the best way to survive and carry on the movement. The period from 1927 to 1932 saw changes in leadership, as well as shifts in tactical orientation. Political debates about the future of the revolution were further complicated by the influence of the International, as well as political developments within the Soviet Union during these years.[43]

At first, surviving CPC leaders sought to strike back at the GMD by launching armed rebellions in two provincial capitals, Nanchang in Jiangxi and Changsha in Hunan. These were areas where peasants' unions were very active, and where popular militias formed. In September 1927 both cities were briefly captured by rebel forces and local communes were declared. But the uprisings were quickly put down by Nationalist military units. Mao, who'd been deeply engaged with rural organizing in the region, and wrote his important "Report on the Peasant Movement in Hunan" in 1926, and Zhu De, a graduate of the Whampoa Military Academy seen as the founder of the Red Army, played leading roles in these rebellions.[44] When the insurrections were suppressed Mao and Zhu led their militia units south to the remote highlands of the area where the borders of Jiangxi, Fujian, and Guangdong provinces meet. There they began establishing rural base areas where the party worked with the local people to create a secure space for political organization and for developing revolutionary policies and practices. Over the next few years, this area grew into the Jiangxi Soviet, one of several such base areas established in different parts of China.[45]

These developments in rural areas were in sharp contrast to the continuing repression of the CPC in the cities. The party maintained and redeveloped its urban organizations, even under conditions of state terror, but had to operate in secret, with a minimal public presence. The central leadership changed hands several times, as arguments over how to continue the revolution and the failure of the autumn 1927 uprisings, led to first one group and then another assuming the top positions. International advisors urged the CPC to maintain its focus on the urban working class as the only revolutionary force in the country, despite the weakness of the industrial sector and the minimal number of workers in the overall economy. The struggle in the Soviet Union between Stalin and Trotsky, involving intense disagreements about the situation in China, also shaped the International's advice and prolonged internal divisions and contention between elements within the Chinese party.[46]

By the early 1930s, the political center of gravity within the communist movement shifted to the rural base areas, especially the Jiangxi Soviet. Mao had been an advocate of working with the peasant farmers as a revolutionary class. He viewed the majority of farmers as agricultural proletarians, especially those who either owned no land and had to sell their labor to survive, or those peasants with only enough land to support themselves and their families. He undertook an in-depth survey of conditions in the rural economy of the Soviet area, carefully delineating the class hierarchy of local society as a prelude to revolutionary transformation.[47]

The CPC experimented with innovative policies to address the inequalities and oppressions of rural society. The soviet government made moderate land reforms to ameliorate the harshest poverty in the area. They also worked to reform the gender system, promoting equality for women in both work and family life. These policies won the party support among the masses of local people and gave activists and cadres experience working with the people to solve economic and social problems embedded in local traditions and culture. The party developed a mass-based military force, the Red Army, a people's army, not a professional corps separate from the masses. These initial efforts at creating a more egalitarian social order became the foundation of later campaigns for land reform and family policies.

The Nationalist government, which continued the Northern Expedition in 1927 and brought most of China under its control by the early 1930s, saw the revolutionary base areas as threats to be eliminated. They feared the popularity of party efforts at land reform and social justice, and were determined to destroy the CPC as a political force. The Jiangxi Soviet became the focal point of their efforts.

The Long March and the Yan'an Base Area

In 1932, 1933, and 1934, the Nationalists conducted encirclement campaigns in which they built a ring of blockhouses surrounding the Jiangxi base area, each year moving closer, aiming to strangle the revolutionary soviet. More than ten million people lived within the soviet territory, and their support for the communist government prevented the GMD armies from seizing the area. But the struggle was costly, and by the summer of 1934 the party leadership considered ways to break the blockade and withdraw from the enclave. In October, a small force launched an attack on Nationalist positions in the northeast, while the main body of the Red Army and the communist militants broke through GMD lines to the southwest. This began what became known as the Long March. Over the next year they walked some five thousand miles, by a circuitous route across the southwest and then into the northwest of China, finally reaching the Yan'an base area in October 1935. Of perhaps 120,000 personnel who left the Jiangxi base, only about 18,000 finally reached Yan'an. Many died in fighting or from disease along the way, while others had to drop out or chose to leave the column as it moved.[48]

Early in the course of the Long March, in January 1935, the main body of the Red Army reached the town of Zunyi in Guizhou province. Here the CPC leadership met, known as the Zunyi Conference, to discuss the recent events that led to the abandonment of the Jiangxi Soviet, the state of revolutionary activity in the country as a whole, and the path going forward. This was a decisive moment in CPC history and the Chinese Revolution. After several shifts in leadership in the

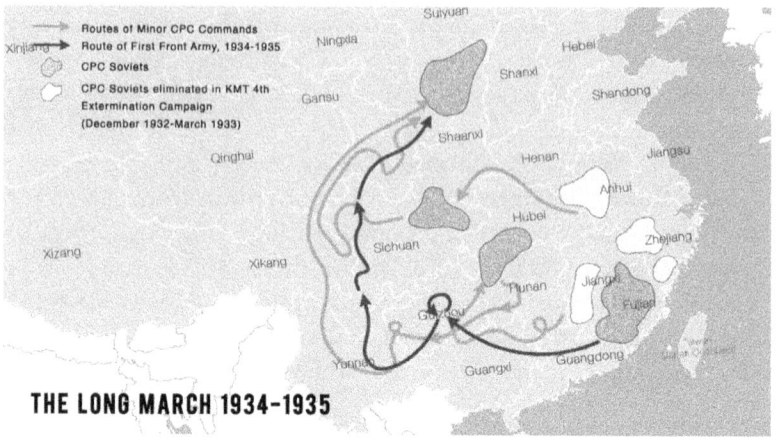

THE LONG MARCH 1934-1935

Route of the Long March (Map: Tina Duong)

years following the end of the First United Front in 1927, the party essentially came to be led by Bo Gu—who had just returned from political study in the Soviet Union—and the International advisor Otto Braun. At Zunyi, the course followed by these two, and by previous leaders like Li Lisan and Qu Qiubai, was criticized. The Political Bureau, the collective's top leadership, decided on a new set of leaders.

While Zhou Enlai was among those criticized for problems that led to the current situation, he was still respected and retained a prominent role. The main shift was that Mao was elected to the Standing Committee of the Political Bureau, bringing him formally into the highest level of decision-making. In practice, Mao became the most influential voice within the leadership, setting the party on a new course. Mao's addition to the Standing Committee brought new stability to the leadership and represented a clear theoretical orientation for the revolution built on the alliance of the rural, agricultural workforce with the small but growing urban industrial proletariat. This applied the general principles of Marxist analysis to the particular material and social conditions of China. Given the long history of commercialized agriculture and the importance of both tenant farming and agricultural day labor within the rural economy, viewing the majority of the peasantry as an agent of revolutionary change made sense, even though it diverged from Marx's classical perception of the French peasantry as an unreliable petit-bourgeois stratum. Mao's advocacy of the peasant line and his leadership over the following years eventually led to his election in 1945 as CPC chairman, a position he retained until his death in 1976.

Once the Zunyi Conference concluded, the Long March resumed, winding through rugged lands in Guizhou and Sichuan, then turned north, crossing through Gansu and into northern Shaanxi province, finally reaching Yan'an in the remote arid highlands of the loess plateau. Along the way the Red Army fought Nationalist forces numerous times, the column was continually harassed by GMD aircraft, and the communists faced severe challenges of weather and terrain. Episodes such as the capture of the Luding Bridge or crossing the snow-covered Jade Dragon Mountains became legendary epic moments in the revolution. To be a survivor of the Long March made one a hero ever after.

On reaching Yan'an, Red Army forces and party activists who had made the journey joined with military and civilian forces already established there, making it the largest and most important of the revolutionary base areas. Yan'an became the headquarters of the CPC and the Red Army, and a laboratory for developing revolutionary politics and culture.[49]

The Yan'an Period, 1936-1945

While Yan'an was the center of the revolution, several important developments took place. Mao undertook his most important theoretical writing during the late 1930s, producing the essays "On Practice" and "On Contradiction," among others. In them, he adapted and articulated the ideas of Marx, Engels, and Lenin within the context of Chinese realities.[50] At the beginning of the 1940s, the CPC undertook a major rectification campaign to raise the political education level among the membership, and seek a shared common understanding of the political tasks and challenges confronting the revolution. Throughout this period the party pursued new initiatives on land reform, economic development, family and gender policy, and other areas.[51]

Another important topic was explored through the Yan'an forums on art and literature. Mao gave a series of talks elaborating on the relationship between the economic and social systems of a society and the cultural and intellectual ideas and practices that arise within a given historical mode of production. These talks emphasized the idea that the revolution was not merely a process of reorganizing the productive economy or redistributing resources, but also must concern itself with creating a revolutionary culture in which the working masses' needs and interests take priority. These ideas later became central to the Cultural Revolution in the late 1960s.[52]

All this occurred within the broader context of Japanese aggression and the continuing conflict with the Nationalists. Japan had pursued its imperialist agenda against China since the late nineteenth century. It took Taiwan after defeating China in a war in 1894-95. In 1910 Japan forcibly annexed Korea, bringing the borders of their empire up against northeastern China. During World War I, Japan made its infamous Twenty-One Demands, seeking economic and political hegemony in China while the main European powers were diverted by their suicidal imperialist war in the trenches of France and Russia. In the 1920s Japan used its concessionary positions in Shandong, Shanghai, and Liaodong to further advance its aggressive agenda. In September 1931 Japanese provocateurs staged a phony attack on the railway line in Mukden (Shenyang) in Manchuria and used this pretext to invade China from their territory in Korea. By 1932 they set up a puppet state called Manchukuo and brought in the deposed emperor of the Qing dynasty, Puyi, as a figurehead. This long campaign of incremental aggression set the stage for all-out invasion as Japan sought to expand its empire and create a Greater East Asian sphere of influence, which they hoped to extend to Southeast Asia and even India.[53]

In December 1936 Chiang Kai-shek flew to Xi'an, the capital of Shaanxi province, where his local commander Zhang Xueliang was based. Zhang, whose warlord father was assassinated by the Japanese in 1928, was deeply concerned about Japanese aggression. He felt that Chiang was ignoring this threat because of his obsession with fighting the communists. Zhang placed Chiang under a kind of house arrest and forced him to negotiate with the CPC about uniting to resist Japanese aggression. Zhou Enlai flew from Yan'an, in northern Shaanxi, to Xi'an and met with Zhang and Chiang. This led to an agreement to form a Second United Front between the GMD and the CPC. This agreement did lead to some cooperation and coordination between the two parties and their military units, but it was a tense alliance, with frequent clashes between the forces, as Chiang maintained his anti-communist views and determination to destroy the revolution.[54]

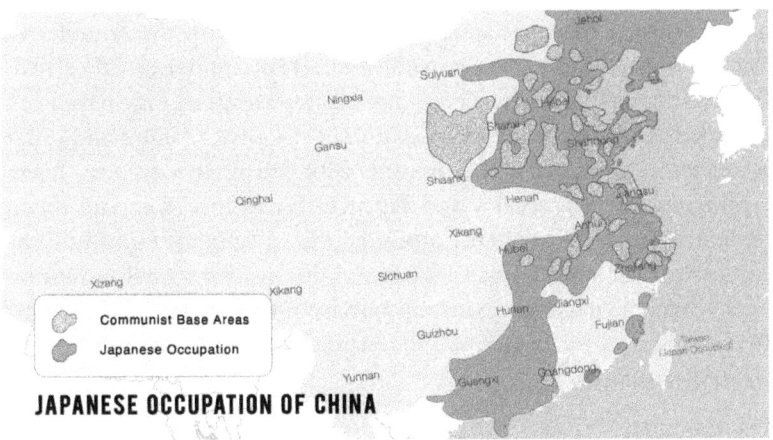

JAPANESE OCCUPATION OF CHINA

Areas of Japanese occupation and CPC-led guerilla resistance (Map: Tina Duong)

The establishment of the Second United Front came just in time, as the Japanese launched their full-scale invasion of China in July 1937. Striking south from Manchuria and west up the Yangzi River from Shanghai, Japanese forces occupied much of northern and central China over the next year. In the course of this invasion Japanese forces committed many atrocities, most notably the Rape of Nanjing in December 1937, when two to three hundred thousand civilians were massacred in an attempt to intimidate and terrorize the Chinese people. This backfired and served to strengthen Chinese popular resistance to the Japanese imperialists. The war continued from 1937 until the final defeat of Japan in 1945. The Red Army conducted guerilla warfare across northern China, in the area of Japanese occupation, tying down more than a million troops and bleeding the Japanese economy while the Pacific war, begun with Japan's attack on Pearl Harbor, Manila, and Malaysia in December 1941, led to further imperialist expansionism and drew the United States into the conflict. The Nationalists withdrew to the city of Chongqing in Sichuan province and carried on a more mainstream war of resistance. Chiang Kai-shek hoarded the military supplies provided by the United States, keeping them in reserve for a later confrontation with the communists.[55]

On August 6 and 9, 1945, the US dropped two atomic bombs on the civilian populations of Hiroshima and Nagasaki, killing more than 150,000 people outright and leading to many more deaths from radiation, cancer, and other effects over the following years. The Japanese had attempted to negotiate an end to the war with the Americans, but the United States was unwilling to accept any terms other than unconditional surrender. The use of atomic weapons may have been intended as much for the intimidation of the Soviet Union and other countries as to shorten the fighting with Japan. In any case, Japan capitulated on August 15, and Japanese forces in China laid down their arms in obedience to the proclamation of Emperor Hirohito. The end of the war with Japan then cleared the way for a final resolution of the struggle between the revolutionary forces of the CPC and the Red Army, and the reactionary government of the Guomindang and its conscript army.[56]

The Civil War

The end of the war was followed by a period of negotiation between the CPC and the Nationalists to create some kind of coalition government in which both parties could share power. Even as the negotiations proceeded, Chiang Kai-shek maneuvered to launch new assaults on Yan'an and a general offensive against the Red Army. By 1946 open hostilities broke out, with the GMD attacking Yan'an and driving the communist forces out of the base area. The Red Army shifted its main base of operations to the Northeast. Soviet military forces invaded Manchuria in the final days of the war, and provided some assistance to the communist army, airlifted units into the region, and turned over captured military equipment. Red Army guerillas attacked Nationalist positions in many parts of the country, but the final triumph of the revolution would be fought out on the plains of the north, the ancient heartland of Chinese civilization.[57]

In the fall of 1948, Red Army troops fought their way southwest out of Manchuria and approached the Great Wall—the old border zone along the northern edge of China proper. By November, Beijing, still known as Beiping, was besieged. Given the great symbolic importance of the city—the capital of the last three imperial dynasties—the communists did not want to wage a pitched battle to take the town. GMD forces were isolated and deprived of supplies. By the end of the year, a negotiated agreement allowed them to be airlifted out of the city, and in January 1949 the Red Army entered Beijing through its southern gates, welcomed by huge crowds along the main street running to the heart of what once again became the capital of a new China.

People's Liberation Army troops entering Beijing, January 1949

As this drama played out in the north, a final battle was being waged in central China along the valley of the Huai River. Known as the Huaihai Campaign, this battle, lasting from November 1948 to January 1949, saw the collapse of the Nationalist army. GMD soldiers were mostly conscripts, poorly trained, and rarely paid. The Red Army, now called the People's Liberation Army (PLA), was just that, a people's army, supported by the masses, with high morale and dedication to the revolution. As the Huaihai battle went on, more Nationalist soldiers threw down their arms and went over to the PLA. The Huaihai Campaign was the beginning of the end for the GMD. Fighting continued through the rest of 1949, and in some areas into 1950, but the Nationalists were forced to retreat, eventually fled the mainland, and took refuge on the island of Taiwan.

The people of Taiwan did not want to be occupied by the Nationalists, and a rebellion against the GMD took place there in February 1947. It was brutally repressed. Taiwan was placed under martial law, which lasted into the 1990s. By late 1949 hundreds of thousands of Nationalist troops were transported to the island, many against their will. Taiwan became the last outpost of the losers in the civil war, protected by the forces of US imperialism. The island province remains under the control of the Nationalists to this day.[58]

The liberation of Beijing in 1949 set the stage for the establishment of the People's Republic, formally proclaimed on October 1, 1949. The long revolutionary struggle to seize power and create a new China was successfully concluded. The CPC was then in a position to lead the Chinese people in building a socialist future.

CHAPTER THREE

BUILDING A NEW CHINA AND THE STRUGGLE BETWEEN TWO LINES: 1949-1976

"The Chinese people have stood up!" With these words, Mao Zedong proclaimed the establishment of the People's Republic of China (PRC) on October 1, 1949. The process of liberation and the extension of the new government's control over all of China's territory, except for Taiwan, continued into 1950. Still, the Communist Party of China (CPC) had already begun to implement new policies in the areas under its control. China faced severe challenges after four years of civil war and nine years of the War of Resistance to Japanese Aggression. Inflation was out of control, millions of people were displaced refugees, and the destruction of infrastructure and industry crippled the economy. As the core of the new system, the CPC, with a membership of roughly a million, confronted the daunting tasks of securing the welfare of and managing the government for a population of 450 million. Several major initiatives were launched to address the needs of the people and to lay the foundations for future development.

Mao Zedong proclaims the founding of the People's Republic, October 1, 1949.

Land Reform

Perhaps the greatest undertaking of the new government was land reform. The redistribution of land from large landlords to individual farming households was done in the liberated zones of the north-east in 1948. As the rest of the country was freed from retreating Nationalists, the stage was set for the fundamental transformation of the agrarian economy. The overthrow of the old imperial system in 1911-12 destroyed the dynastic state and was followed by efforts of the urban bourgeoisie to create a republican system. But it largely left the old landowning elite in power in the countryside, where the vast majority of the people lived. Despite some rural reforms in the 1930s, the relations of power remained much as they had been for centuries. Land reform finally broke the power of the landlords, creating an equitable distribution of assets and resources for hundreds of millions of Chinese farmers. Ending the old system of land tenure was essential for the development of a new socialist agricultural economy.[59]

Land reform followed a careful and deliberate course, relying on the active participation of the peasants in villages rather than being proclaimed and imposed from the outside. Small teams of CPC cadres went to villages to observe and analyze the particular conditions of local society. Over a period of weeks, they talked with people about their lives in the old society, about their suffering at the hands of the wealthy and powerful landlords, in a process called "speaking bitterness." Meetings would be held for people to share their experiences and express their anger at the landlords. Eventually, mass meetings were held where landlords were confronted by the people, criticized and denounced for the exploitation and oppression they'd inflicted, and the privileges they'd enjoyed at the expense of the peasants. In the end, old land titles and rent or tenancy contracts were publicly burned, and new deeds were drawn up, giving each adult title to enough land to support themself. Landlords were allowed to retain small plots for their own use, like everyone else.[60]

These meetings sometimes erupted into violence against landlords. The rage built up over generations led to beatings and other physical attacks on especially cruel or vicious landlords. Some were killed, though the great majority, after having been struggled against by the people, assumed new roles as small farmers on remnants of their former estates. Land reform was a cathartic process, in the course of which old bonds of obligation and subservience within village society were broken. It was a political process, and also a psychological one. It liberated the economic potential of the countryside, as well as peasants' consciousness, fostering a sense of their power to create a better future.

The Marriage Law of 1950

In 1923, radical writer Lu Xun gave a speech at the Beijing Women's Normal College called, "What Happens After Nora Leaves Home?" Henrik Ibsen's play, *A Doll's House*, was performed in China a few years earlier, setting off a lively discussion of women's issues. The protagonist of the play, Nora, struggles against the constraints of her bourgeois marriage, finally fleeing her home for an uncertain future. While many in China viewed the story as one of a woman taking control of her life, Lu Xun raised the question of Nora's economic prospects. She had no real skills for the workplace, and was likely to either become a parasite on others or wind up falling into prostitution or some other form of subservience to survive. Lu Xun's speech was a call for women's economic and political empowerment within a broader system of social production and distribution.[61]

In 1950, one of the first legislative acts of the new PRC government was the Marriage Law. It prohibited the old practices of arranged marriages in which the feelings or needs of individuals were subordinated to the economic interests of clans that often effectively functioned as corporate units. Marriage was now a contractual agreement between two consenting adults, either of whom could subsequently end the marriage. The CPC had long maintained a commitment to women's rights and reform of the gender system. In practice, the party often fell short of its goals: the Red Army remained a largely male fighting force, and the party leadership was almost entirely men. Nonetheless, the Marriage Law was a dramatic step toward equality of the sexes.

One provision is noteworthy, bearing in mind Lu Xun's concerns about women's economic position. The land reform process going on simultaneously, gave women title to land of their own when redistribution took place in villages across China. Rather than allocating land only to households, normally headed by men, every adult member of a household received their own property. This gave women the economic basis for an independent life if they should choose to leave a marriage. There were many divorces in the early 1950s as both men and women ended relationships that might have been imposed on them against their feelings, and then sought partners of their choosing. Almost everyone wound up married, but now those links were voluntary.[62]

The role of women in Chinese society, in rural villages and urban workplaces, underwent significant changes after liberation. This was not an instantaneous, nor a complete transformation. But women played major roles in production, took part in new areas of employment, gained a larger presence in the People's Liberation Army (PLA), and won new educational opportunities. The slogan "Women hold up half the sky" encapsulated the ideals of gender equality that were the policy on women's issues for the party and the government. The All-China Women's Federation became the institutional representative of women across the country, publishing magazines and newspapers, holding conferences, and articulating their interests and concerns.

The Sino-Soviet Friendship Treaty

Land reform and the gender system were important domestic issues of the new government, but China's development was also strongly shaped by the relationship between the revolutionary government in Beijing and the Soviet Union. China's revolutionary struggle was supported by the Communist International and by the Soviet government, but Soviet leader Joseph Stalin continued assisting Nationalist authorities until as late as 1948. The Soviet Communist Party (CPSU) maintained the view of the Guomindang (GMD) as the representative of a national bourgeoisie that would develop the country before a true socialist revolution was possible. Stalin never fully embraced the Chinese model of relying on the peasants as agricultural proletarians, allies of the small industrial working class. Only when the conflict between the Red Army and the collapsing GMD was clearly in its final phase did the Soviets finally accept the CPC's successful strategy.

Shortly after the People's Republic of China was established in October 1949, Mao took his first and only journey outside China. He traveled to Moscow with a large entourage, and met with Stalin and other Soviet leaders to negotiate a treaty of friendship and mutual assistance. The negotiations were not without their challenges, but in February 1950 a treaty was signed, and Mao returned to Beijing. This treaty became the foundation of a period of significant cooperation between the Soviet and Chinese governments and their communist parties. The Soviets were to provide loans (at interest), technical assistance in the form of advisers, engineers, scientists, etc., and equipment, to help China begin the process of modern industrialization. This aid was vital to China's initial phase of socialist development, but had its complications. As seen later in this chapter, tensions emerged about how Soviet and Chinese approaches to industrial management, agricultural collectivization, and military organization diverged. These resulted in the 1959 Sino-Soviet Split.[63]

The Korean War

The challenges facing the Chinese people after liberation were made even more complex by the outbreak of war on the Korean peninsula in June 1950, less than ten months after the PRC was established. Korea and China had a long history of close ties, both cultural and political. Japan annexed Korea in 1910, abolished the monarchy in Seoul and made Korea a colony of the Japanese empire. When the Pacific war ended in 1945 Japan's territory was rolled back to its pre-1898 borders, and Korea was to be independent again. As the war drew to a close in

August, Soviet forces entered the northern part of the peninsula while American troops occupied the south. A division of the country, meant to be temporary and similar to that imposed on Germany, settled in as political administrations were established in both sectors. In the south, US occupation forces suppressed socialist and communist political movements as well as trade unions. And, violating the agreement to pursue reunification of the peninsula, installed a government led by Syngman Rhee, a Presbyterian minister who had lived outside Korea since 1896, mostly in the US. After the 1948 Soviet withdrawal from the north, the Democratic People's Republic of Korea (DPRK) was formed. It was based on anti-Japanese communist resistance forces led by Kim Il-sung, that fought the Japanese colonial government from bases on the border with China since the 1920s and had widespread popular support.

By the end of the 1940s the political situation deteriorated. Under US domination and during the Cold War, South Korea (ROK) adopted a strongly anti-communist posture. The Cold War context reinforced American hostility as well. Tensions deepened, both sides viewed the other as a threat, and by the summer of 1950 they reached a breaking point. North Korean forces, anticipating an attack by the ROK, moved south across the line of demarcation. Over the following months, the war developed first in favor of the north, which advanced almost to the southern tip of the peninsula and was on the verge of victory. Then, American-led forces under a United Nations fig leaf of legitimacy intervened and pushed back the DPRK forces. By October, US forces were close to the Chinese border. The PRC sought to defuse the situation, reaching out to Washington via its embassy in India, but the Americans did not respond. In early November six hundred thousand Chinese People's Volunteers soldiers crossed the Yalu River and joined the fight to defend the DPRK. Before long, fighting reached a stalemate near the original line of demarcation. In mid-1953, an armistice was agreed upon. It ended the fighting but left a state of war technically in place, one persisting to the present day.

The War of Resistance to American Aggression, as it is known in China, was a serious drain of resources and manpower in the first years of the PRC. It was an existential threat to the new government, as US forces threatened to invade and aircraft dropped bombs in northeast China. Clandestine operations were carried out by the

US, including air raids on Shanghai and sabotage in Fujian province, across from Taiwan. The war also created economic problems, both in terms of materials and supplies used for economic development being consumed in the war effort, and as concerns about corruption and profiteering emerged. These were part of a larger set of issues involved in transforming China's urban economy, the initial steps towards creating a socialist industrial system.

Urban Transformations

Although China's revolution had been largely carried forward by mobilizing the peasantry, the goal was a modern, industrial, and socialist economy, as part of a new, just, and equitable society based on the shared wealth produced by the workers. This was understood to be a long-term project, one which would take decades or even a century to be fulfilled. The first stage involved stabilizing the urban economy and creating state-owned enterprises and other forms of public property in the means of production. The political framework was called the New Democracy.

Mao enunciated the ideas of New Democracy in the late 1940s. The concept built around existing social and political conditions in China. Although many patriotic capitalists were wanting to be part of the new China, there were also political elements outside the party, such as professionals, intellectuals, and others, who wanted to help develop the country. New Democracy brought these groups into the political system, primarily through the Chinese People's Political Consultative Conference (CPPCC). This body would offer advice to the new government, reflecting the particular interests and concerns of its constituents.

To develop China's industrial economy, the CPC and the PRC government began a gradual process of nationalizing existing productive and financial enterprises. This was done through incremental shifts from private to public ownership and management. At first, private owners were allowed to retain control over a business under government supervision. Through the early 1950s, new personnel were added to management to learn the details of operating and administering the enterprise. The private owners were bought out over a few years, though many retained leading roles in operating the enterprises. By the mid-1950s, most productive units were brought into full state and public ownership. Former owners either continued in their management or retired with government compensation.

Nationalizing existing businesses was only one part of developing urban industry. Investment in growing productive capacity came from two sources. One was surplus accumulation from advancements in agricultural production. The other was aid and investment from the Soviet Union. Agricultural cooperative development and collectivization discussed below, generated rising surpluses through 1958. This provided significant revenues through sales on the international grain markets, which were then used to purchase industrial equipment and materials. Soviet aid provided loans that could be used for such purchases, as well as direct transfers of machinery and materials, and deployment of technical advisers to help with the construction and operation of new factories and other productive activities. By the end of the 1950s a solid foundation of modern industrial production was created in cities and towns, especially in the east and northeast of the country.[64]

The process of urban transformation had its challenges. As noted above, during the war in Korea and the initial phases of nationalization and development of the urban economy, contradictions developed that undermined building a new economy and diverted scarce resources and funds into private corruption. This included profiteering on military supplies. In other cases it was market manipulation and abuse of power by government or party officials draining public resources for private gains. These corrupt practices became the target of political campaigns, to ferret out venal officials and businessmen and promote public oversight. The Three-Anti and Five-Anti Campaigns in 1951 and 1952 were aimed at bringing bribery, waste, bureaucratic obstruction, tax evasion, and outright theft of public property under control. While corruption remained a concern in later years, and revived in the period of reform and opening after 1979, by the mid-1950s these campaigns significantly curtailed the worst abuses of power and the actions of profiteers.[65]

Agricultural Collectivization

The agricultural sector in China's economy was composed of equalized small holdings, with village households making the production decisions. Productivity increases were modest, with most households meeting their own needs and generating a small surplus for saving and investing in improvements to their farms without major efforts. To enhance productivity, to yield growing surpluses to both sustain urban populations and provide grain for export (to bring in funds for further industrial development), larger economies of scale were needed. Traditional rural landholdings were scattered, small strips of land, and weren't suited to making further advances in the rural economy. Party and government leaders saw agricultural collectivization as the best path toward greater productivity and the development of the socialist economy.

Given the social nature of the Chinese Revolution, with the peasant masses as the mainstay of the liberation struggle, collectivization was pursued by demonstrating to farmers that economies of scale gained by combining smaller units yielded benefits for primary producers on the land and for the overall project of socialist development.

While collectivization carried out in the 1920s and 1930s was contentious in the Soviet Union, with widespread resistance by elements in the countryside, in China it was carried out in a step-by-step process from the mid-to-late 1950s, with gains made at each stage fostering steps toward larger units.

At first, small mutual aid teams formed within villages, a few households sharing assets and resources, including tools, farm animals, and labor. These were followed by lower-level agricultural producers' cooperatives (APCs), which might bring together a whole village, and then by higher-level APCs that included a few villages. Finally, agricultural collectives formed, with individual households merging their land titles into collective units, with shared administrative and decision-making powers. By fall of 1958 this process culminated in the formation, in Henan and Shandong provinces, of the first People's Communes. These were large-scale collectives encompassing numerous villages, roughly equivalent to the size of a county in rural America. Mao, on an inspection visit to one of the new communes, made the comment that "People's communes are good!" This was picked up by reporters traveling with the chairman, and soon appeared in headlines across China, triggering a great surge in commune formation, leading to the launching of the Great Leap Forward.

The Great Leap Forward was a radical initiative to accelerate economic development. It mobilized the enthusiasm of the masses in the countryside to achieve higher levels of production, and develop new

Poster showing the ideal image of a people's commune, 1958

forms of social life in the communes. Experiments in collective food preparation and dining, child care, and other kinds of social provision normally done on a household basis were undertaken, both to foster greater social solidarity and free labor for application in the fields. Efforts were made to overcome the division between urban and rural livelihoods, by developing small-scale local light industries, and even producing heavier products like iron and steel based on decentralized local initiatives. The Great Leap Forward also saw the mobilization of large-scale construction projects for infrastructure, especially for water management, creating reservoirs, canals, irrigation systems, and flood control works.

Many aspects of the Great Leap Forward were successful, especially infrastructure development. But it was plagued by serious problems resulting in a food supply crisis in rural areas by the spring of 1959. The causes are discussed below, but misallocations of food, especially grain, led to drastic supply shortfalls in the villages. In some areas people died from malnutrition or disease and other health issues made worse by lack of adequate food. Several million people died, above the rates normally expected. It took time for the seriousness of the crisis in the countryside to be understood, but by the summer of 1959, the party began to address the problems and sought solutions. This was a major crisis for the CPC and led to serious adjustments in policy and leadership. Before discussing these, it is necessary to fill in the background of what is known as the Struggle Between Two Lines.[66]

The Struggle Between Two Lines

As the leadership of the CPC and the PRC government embarked on building a new socialist China in the 1950s, there was broad consensus that the revolution's mission was to develop a just, equitable society in which the means of production were social property. The fruits of the workers' labor, urban and rural, were to be shared among those who had produced them, with necessary accommodations for those unable to contribute to productive tasks. A modern socialist industrial economy was the basis for this new order, and construction of new productive capacity was the immediate task at hand.

There were, however, significantly differing views among the leadership as to how this mission of building a socialist China should be pursued. As the decade advanced, the differences coalesced into two increasingly antithetical camps, and the debates and conflicts between them became known as the Struggle Between Two Lines. One perspective, associated with Mao, having wide support within the party and broader society, relied on mass mobilization to drive the developmental process. Mao and his supporters believed that the enthusiasm of the fully mobilized masses, could become a surrogate form of capital, and bring about the rapid expansion of production and improvement in people's livelihoods, the keys to a socialist future. The other approach to development, associated with Liu Shaoqi and Deng Xiaoping, argued that forging a modern economy was technically challenging. Those with the skills, talents, training, and expertise to carry out the requisite tasks should be entrusted with that work and allowed to carry it out efficiently and effectively. This may be thought of as a more technocratic model of development.

The divisions between these two positions within the party were manifested in several ways, from the management of industrial enterprises to the process of agricultural collectivization, and to the operational organization of the People's Liberation Army (PLA). These differences were also influenced by the Soviet Union's own developmental experience and advisers. In industry, for example, the Soviets advocated a system called One-Man Management, a system in which a leader oversaw the operations of an enterprise in a top-down hierarchy, to promote efficiency and increase productivity. Mobilizationists, on the other hand, promoted worker participation in decision-making, and their knowledge and experience as a great source of better practices in running factories, mines, or other productive activities. The Soviet experience in agricultural collectivization was a very top-down process, while China's was a more organic, incremental, and village-based process. In military affairs, the Soviet Red Army was highly structured with a sharply differentiated system of ranks and insignia. The Chinese Red Army, and then the PLA, emphasized leadership from the ranks and a more collective form of command and control.

The tensions between the two lines grew stronger as urban and rural development advanced. They reached a crisis when the problems of the Great Leap Forward forced the party leadership to confront these issues head-on. The differing positions on socialist construction were also embedded in a context of increasing bureaucratization within the party and the government. The CPC grew from a membership of around one million in 1949 to over ten million by the late 1950s. Many former Nationalist officials and other employees of private enterprises or financial institutions joined the party. Some joined out of sincere patriotic wishes to be part of creating a new China, but others joined for reasons of personal advancement and careerist opportunism. The CPC, the key institution guiding the country's development, found itself in a position to make critical decisions about social, economic, and political life. Large complex organizations, as the sociology of institutions has shown, tend to become self-referential and self-protecting. These bureaucratic tendencies were evident in the CPC as the first decade of the PRC passed. Technocratic leanings of some party members gave rise to certain attitudes that began to separate the party from the masses. The combination of the Struggle Between Two Lines and creeping bureaucratization of the CPC were major contributors to the failures of the Great Leap Forward. In August 1959 the party leadership gathered at Lushan, in central China, for a meeting to discuss the country's crisis and how to resolve it.[67]

The Lushan Plenum

By the time China's leaders assembled at Lushan, the crisis was becoming better understood. Several factors combined to produce the drastic shortfall of food supplies in the countryside. Perhaps the most important was the distortion of harvest figures for the fall of 1958 and the spring of 1959. Many cadres in communes across the country had made slightly upward exaggerations in reporting yields from their fields. These were then further inflated modestly as they were transmitted up the hierarchy, with the cumulative result that the central authorities were working with numbers overstating the available grain supply. This led to miscalculations when making decisions about procurements for the cities and the international market, and establishing targets for the coming season. This led to too much grain being taken from the communes, leaving too little to meet the needs of the people for food and for seeds for the coming planting season. Cadres who made these exaggerated reports sought to cover their tracks and avoid responsibility, which delayed a clear understanding of the gravity of the situation and in organizing an effective response.

These failures of the administrative system were exacerbated by two other serious developments. One was natural: the weather, remarkably mild through much of the 1950s, turned bad in 1959, so the steady harvest growth since land reform now turned into a lower-than-expected yield. This was concealed by the falsified reports sent up by cadres in many locations. The second was political. The Soviets became increasingly critical of China's developmental process, the leadership of the Soviet Communist Party viewing it as adventurist and out of control. They abruptly withdrew their advisers and canceled their aid projects in the spring of 1959, much to the consternation of many Soviet comrades working closely with their Chinese counterparts. This caused major problems in urban areas and put further pressure on the food supply, since the Soviets demanded payment in goods rather than Chinese currency for many of the loans extended under the Treaty of Friendship. The combined effect was the food crisis in the countryside and the delayed response by the party and government leadership.

The relationship between China and the Soviet Union was also strained by diverging views of the international situation. The USSR, under the leadership of Nikita Khrushchev, promoted a line of "peaceful coexistence" with Western imperialism while China endorsed revolutionary movements in many countries around the world. Mao and others in the Chinese leadership increasingly viewed the Soviets as "revisionist." These tensions deepened in the course of the 1960s.

When the Lushan meeting convened, the CPC leadership was divided over questions of why the crisis developed and how it should be dealt with. These divisions reflected the Struggle Between Two Lines. The advocates of mobilization, grouped around Mao, criticized the bureaucratic self-interest and defensiveness of cadres who distorted harvest figures, while the technocrats, aligned with Liu Shaoqi and others, blamed the decentralized and, in their view, excessive reliance on enthusiasm for producing a chaotic situation where oversight and management were rendered ineffective.

A confrontation developed between Mao and Peng Dehuai, the minister of defense.[68] Peng drafted a letter criticizing Mao and the mass mobilization policies, which he circulated among the leadership before delivering it to the chairman. Mao viewed this as a divisive factional move and criticized Peng for his actions. After lengthy discussions, the plenum reached several decisions. In dealing with the food crisis, the strongly centralized organization of the communes was to be eased, giving individual households the option of planting small gardens to grow vegetables and other food to supplement the output of collective farming. Rural markets would be expanded to promote a more efficient allocation of supplies and alleviate shortfalls. These measures were implemented from the fall of 1959 onward and revived production and distribution to end the food crisis. By 1961, grain production returned to 1958 levels, and continued to grow.

The plenum also made key political decisions. Peng was removed as minister of defense. He was later placed in charge of industrial development in the southwest provinces. Mao was called upon to step down as president of the PRC, and was succeeded in this role by Liu Shaoqi. Mao was also directed to step back from the day-to-day oversight of party affairs. He retained his position as chairman of the CPC, but was urged to devote his attention to more theoretical work. These measures reflected continuing tensions between the two main groupings within the party, with concessions made by both sides. But it left the contradiction between their positions unresolved.[69]

The Socialist Education Movement

Mao's concerns about bureaucratic tendencies within the CPC were grounded in his analysis of the experience of the CPSU and the history of the USSR. In the late 1950s, as problems with the Great Leap Forward were developing, Mao wrote a set of reading notes on the Soviet book *Political Economy: A Textbook* and a couple of critical essays on Stalin's *Economic Problems of Socialism in the USSR*, which he drafted as a review of the political economy text.[70] In these, he argued that the Soviet party became alienated from the masses of the people and developed policies distorting the process of socialist development in top-down ways. Soviet economics under Stalin, and more so under Khrushchev, operated in an exploitative manner, rather than empowering the working class. Mao feared bureaucratic elitism could emerge within the Chinese party and threaten the whole process of socialist development.

By 1962 Mao was determined to make another effort to criticize the sometimes high-handed conduct of cadres, especially in the countryside, where the great majority of the people still lived. He developed a plan for a Socialist Education Movement to send higher-level party leaders to villages in different parts of the country to investigate the relationship between local cadres and the masses. Young people from the cities were urged to visit rural areas during school vacations to learn about the conditions of life for the peasants. Mao's goals were to bring party cadres back into a closer relationship with the people and to awaken revolutionary consciousness among young people who had grown up since liberation.

The Socialist Education Movement ran into difficulties from the start. Many central government leaders were not interested in going to villages to spend time this way. When officials did visit, local cadres sought to divert attention from themselves by blaming problems on former landlords and their families or on rightist elements who were criticized during the 1957 Anti-Rightist Campaign.[71] This tactic, known as "turning aside the spearpoint" would be seen again during the Cultural Revolution, discussed below.

The Socialist Education Movement did have some positive outcomes, mostly giving some urban youth a greater awareness of conditions in rural China. This generated a feeling among many of them that the promises of the revolution remained somewhat unfulfilled in the villages, where incomes and material standards of living were well below those in the cities. This consciousness became an important element in the emergence of young activists when the Cultural Revolution erupted as a mass movement in 1966. Mao's frustration with the resistance to the Socialist Education Movement deepened his resolve to reinvigorate the links between the party and the masses. By the fall of 1965 he began what was to be his final struggle for this objective.

The Great Proletarian Cultural Revolution

In January 1961 a play by the historian Wu Han was put on at the Beijing Capital Theatre called *Hai Rui Dismissed from Office*. The story concerned a Ming dynasty official who was relieved of his official position in a conflict with the emperor over agricultural policy. The policy later proved to be a failure, and Hai Rui was reinstated and recognized as a good official. While the play was supposedly just telling this historical tale, there was a long Chinese theatrical tradition of using events from long ago to comment on current affairs. Wu Han's play was widely understood to be a criticism of Mao and a defense of Peng in the context of the Great Leap Forward and the events of the Lushan plenum discussed above.[72]

While the play's historical dimensions were discussed in academic circles, the larger political implications were not publicly debated. In the summer of 1965, Mao, already concerned about being marginalized from day-to-day political activity, sought to initiate a critique of Wu Han's position. But he couldn't get an article about this published in Beijing's main newspapers. He withdrew to Shanghai, and in November an essay criticizing Wu Han appeared in a local paper under the name of Yao Wenyuan, one of Mao's personal staff members. This was the first public action in what became the Cultural Revolution.

Mao's initial objective was to reopen debate about bureaucratic tendencies in the party, and the relationship between the CPC and the masses. He saw the cultural sphere, where many books, films, plays, and other creative productions still expressed ideas and attitudes derived from prerevolutionary society instead of the new socialist culture emerging in economic and social life, as a critical arena for confronting these issues. Through the winter of 1965-66, discussion of cultural issues and the nature of the party developed, but were

Red Guard rally, Beijing, 1966

often limited in scope, and CPC leadership figures sought to contain the conversation within academic bounds. Mao, on the other hand, wanted a broad public debate and pushed to have people outside the party included in the process.

By May things heated up. Mao and other leaders associated with him sent work teams to university campuses in Beijing, where some students and junior faculty attempted public critiques of bureaucratic elitism among university administrators. At first, party secretaries on campuses tried to divert the work teams' attention toward old rightist elements, but when a "big character poster" was put up attacking the leadership at Beijing University, Mao declared his support for the rebels, and the phase of mass participation in the Cultural Revolution got underway. Mao visited campuses in the northwest part of the city and put up his own poster, saying it was right to rebel against reactionaries, and calling on people to "bombard the headquarters."[73]

In August, the first Red Guard rallies were held. These carried on through the fall and brought millions of young people to Tiananmen Square, where they were greeted by Mao and other leaders of the struggle. The development of the Red Guard movement was complex, with different groups organizing competing factions that sometimes led to outright clashes. Red Guards traveled around the country to visit revolutionary history sites and learn from older comrades. This was facilitated by granting free travel on the railroads to Red Guards. Many visited Jingganshan, the first revolutionary base area, in the mountains of Jiangxi province. Others went to Yan'an or other sites along the route of the Long March.[74]

Not only students were taking up the cause of the Cultural Revolution. Many workers, especially younger ones, took part in political activities. Shanghai, the country's most populous city, was a significant arena for this. It had many factories as well as docklands and warehouses shipping goods into and out of China. Through the fall of 1966, several groups of workers emerged and raised industrial production issues such as factory management, wage rates, and safety concerns. There were tensions and confrontations between workers and management of factories and other economic units, which included criticisms of party personnel in the factories. In November a group of workers attempting to go to Beijing by train was blocked by local party authorities. This deepened the rift between the workers and the CPC.

By January 1967 the groups coalesced into two large mass organizations. They presented various demands to the Municipal Party Committee, but the bureaucratic elements in Shanghai resisted giving in to the mass organizations. The result was a huge mass rally that brought together both major workers' groups on February 5. The Municipal Party Committee was overthrown and workers established what they called the Shanghai Commune, emulating the experience of the 1871 Paris Commune, which had been the subject of many study groups over the previous months. A leadership committee was elected and the commune became the effective government of Shanghai.[75]

This proved to be a critical turning point in the Cultural Revolution. Commune delegates, including Mao's former personal staff member Yao Wenyuan, Shanghai CPC leader Zhang Chunqiao, and a newly rising factory representative Wang Hongwen, were sent to Beijing to meet with Mao and the Cultural Revolution group leaders. In their talks, Mao praised the workers' mass action, but argued that it was not proper to overthrow the party, the revolutionary project needed it as its leading and guiding organization. The point, in Mao's view, was to bring the party back into a close relationship with the masses, not do away with the CPC altogether. After several days of talks, the delegates returned to Shanghai.

At the end of February, implementing Mao's guiding instructions, the commune voted to dissolve itself as the sole political body in the city, and a new leading organization was created, the Revolutionary Three-in-One Committee. This included representatives of the CPC, the PLA, and the mass workers' organization. The idea was to reinvigorate the links between the party and the masses. This new form, the Revolutionary Committee, became the standard institutional expression of the political resolution of the Cultural Revolution's basic goal of shaking up the party, especially the leadership, and forging renewed bonds between the CPC and the working class. Over the next two years Revolutionary Committees were established in provinces, cities, and counties across the country.

As this process progressed, serious tensions and conflicts still emerged in many places. Some were political divisions based on rival interpretations of Mao's thoughts and other revolutionary texts. Others were expressions of institutional interests by competing groups within the economy or the government. And some were instances of personal ambition or score-settling largely devoid of political content. The Cultural Revolution was sometimes chaotic, as mass struggles often are, not a clearly organized and carefully managed process. In Wuhan during the summer of 1967, this even extended to rival factions within the People's Liberation Army backing contending workers' organizations in the large industrial city. This nearly descended into civil war, and required Zhou Enlai's intervention to defuse the situation. A year later, Red Guards were "sent down" to villages across the country to learn from the lives of the peasants and bring the turmoil of Red Guard factionalism to an end.[76]

The rustication of millions of educated youth to rural areas yielded some of the most positive results of the Cultural Revolution. While they did not become the most productive agricultural workers, many urban young people became teachers, ad hoc engineers, or "barefoot doctors," bringing minimal but vital health care to farming communities, benefiting many people in the countryside directly from the skills and talents of the Red Guards.[77]

In April 1969 the CPC convened its Ninth Party Congress. This was meant to mark the end of the mass-struggle phase of the Cultural Revolution. Revolutionary Committees had been established everywhere across China, and it was time for the party to reconstitute itself as the leading force in political life. A new central committee was elected with the highest proportion of new members of any previous congress. Many were from the PLA, giving the military its greatest representation ever. A new party constitution was adopted, with Lin Biao named as Mao's "close comrade in arms and successor" in a gesture aimed at ensuring a stable transition in the event of Mao's passing. But this proved not to be the way things would turn out.

Developments in the 1970s

While the Ninth Party Congress was getting underway, tensions with the Soviet Union, rising throughout the 1960s, reached a crisis point. In March there was fighting between Soviet and Chinese troops along the Ussuri River that formed the border between the two countries in Heilongjiang, in the far northeast of China, and on the frontier between Xinjiang and Soviet Central Asia. These conflicts generated fear that an all-out war between the two countries was a real possibility. This led to a large-scale program to build bomb shelters in Beijing and other northern cities. It also resulted in changes in Mao and other leaders' strategic thinking, reconfiguring the geopolitical landscape of China's foreign policies. The Soviet Union was no longer seen as a fraternal socialist state with some different features from those of China, but was now critiqued as embodying "social imperialism," meaning socialist in form but imperialist in actions. This new assessment led to two dramatic developments at the beginning of the 1970s. First was the fall of Lin Biao, and second, was the opening to the United States.

Details of the events that took place in the late summer of 1971 remain unclear, but Lin Biao seems to have opposed the new line on the Soviet Union and the new perspective on global affairs that it shaped. Up to the end of the 1960s, China's primary contradiction in the world was the clash between the socialist camp and the imperialists, led by the United States. But now Mao and others felt the Soviet Union was the greater threat to China, and the US was a spent force after it lost its imperialist war in Vietnam and faced significant political unrest at home. Lin Biao, and perhaps others in the military leadership, apparently rejected this analysis and viewed Mao as leading the country in the wrong direction. The exact course of events remains unclear, but in early September a plane carrying Lin and several

Nixon, arriving in Beijing, immediately shakes Zhou Enlai's hand. (Photo: White House)

members of his family and other PLA members, apparently bound for the Soviet Union, went down in Mongolia, killing all aboard.

The shift in the analysis of the international situation and development of the idea that the principal contradiction facing China was no longer with American imperialism, but rather with the Soviet Union, led to the 1971 opening to the United States. The Chinese reached out to an American ping-pong team taking part in the world championships in Japan and invited them to come to Beijing for some friendly games. The Americans agreed, with the Nixon administration's approval. This created a small moment of positive media coverage in the West. In the fall of 1971, Nixon's top advisor Henry Kissinger made a secret visit to Beijing, during which plans for a later trip by President Nixon were agreed upon. That trip took place in February 1972. Nixon met with Mao Zedong, Zhou Enlai, and other leaders. At the end of the two-week visit, both sides issued the Shanghai Communique, establishing the fundamental principles for the relationship between the two countries. These were based on mutual respect for the sovereignty and territorial integrity of each side. The United States acknowledged that all Chinese on both sides of the Taiwan Strait agreed there is one China, Taiwan is part of China, and the American government did not dispute this. The United States recognized the People's Republic as the sole legitimate government of China.[78] This process unfolded over the following seven years with final full recognition implemented in January 1979.

Through the early 1970s there continued to be serious political conflicts within the CPC. The mass participation phase of the Cultural Revolution concluded with the Ninth Party Congress, but issues of cultural policy and bureaucracy remained contentious. Activists associated with Mao, including his wife Jiang Qing, Yao Wenyuan, Zhang Chunqiao, and Wang Hongwen, and other former members of the Cultural Revolution Group, continued criticizing Deng Xiaoping and other party leaders viewed as promoting policies that undermined links between the party and the masses. Deng was removed from leadership during the Cultural Revolution, but returned to a top role guiding economic and scientific policies in 1971. The conflict between these positions within the CPC continued through the mid-1970s.

In the history of the People's Republic and the Communist Party, 1976 proved to be a critical year. In January, Zhou Enlai died. In April, internal struggles within the CPC led to a mass incident in Beijing when people honoring Zhou on the Qingming Festival clashed with police in Tiananmen Square. Contradictions in the Party were quite complex at this time, but no clear resolution of the issues had yet emerged. In July, Zhu De died, a senior leader and the founder of the Red Army in the late 1920s. A massive earthquake in the city of Tangshan, east of Beijing, killed nearly three hundred thousand people later that month. Finally, on September 9 Mao died.

Mao's death was quickly followed by the suppression of the group led by Mao's wife, Jiang Qing, referred to as the "Gang of Four" by the now-dominant leadership. In early October, Jiang, Yao, Zhang, and Wang were arrested by security forces. Over the ensuing two years, there was a political debate within the party about how to proceed with China's economic and political development. The details remain obscure, but the outcome became clear by the end of 1978. The Struggle Between Two Lines, that animated China's political life over the previous three decades, was agreed to be resolved. The party remained the leading force in the process of building socialism with a commitment to maintain close links with the masses and take the needs and interests of the people as the guiding basis for the future. How this would be carried on remained a matter of discussion, and the lessons of the previous period needed to be analyzed and understood.

As the 1980s got underway, with Deng Xiaoping now firmly in place as the principal leader, new policies were formulated with the goal of accelerating China's economic growth and reorienting the process of development.[79]

CHAPTER FOUR

REFORM AND OPENING UNDER DENG XIAOPING

Into the Reform Era

With the death of Chairman Mao in September 1976 and the subsequent arrest of Jiang Qing, Yao Wenyuan, Zhang Chunqiao, and Wang Hongwen in October, the CPC and the PRC government entered a period of political transition. This led to the reemergence of Deng Xiaoping as the effective leader of both. From the autumn of 1976 until November 1978, Deng remained largely out of public sight, while debates and discussions within the party sought to resolve long-standing disagreements over how best to develop China's modern socialist economy. These involved the assessment of Mao's historical role and determination of the overall policy orientation for future development. In December 1978, at the third plenum of the Eleventh Central Committee meeting in Beijing, Deng consolidated his position, becoming general secretary of the party and a vice-premier in the government. Hua Guofeng, who succeeded Mao as party chairman, was marginalized and gave up his leading positions over the course of 1979 and 1980.

Deng Xiaoping's leadership put China on a new course. While the PRC achieved significant progress over its first thirty years, with average annual growth in the economy of 3-3.5 percent, and significant improvements in the material conditions of life for the people, this development only slightly exceeded the growth of the population.

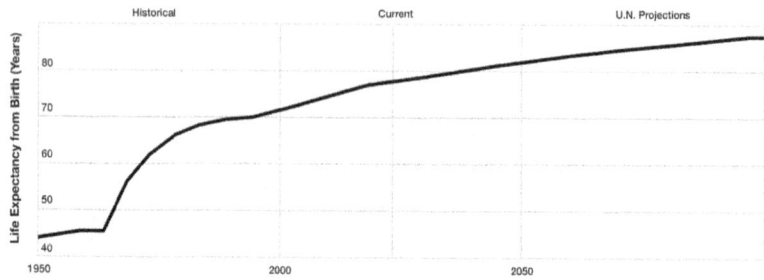

Extension of life expectancy since liberation (Chart: Hannah Craig)

Life expectancy was dramatically extended, infant mortality greatly reduced, and improvements were made in housing, health care provision, and education. Basic infrastructure was developed and core sectors of heavy industry were prioritized, along with the enhancement of agricultural production. It was based on these accomplishments that the next phase of economic expansion was launched. But the goal of socialist construction, attaining a level of prosperity to realize the distribution of social wealth on the principle of from each according to their ability, to each according to their work, was still out of reach. Socialism, and eventually communism, would be an economy not of shared poverty, but of abundance.

The basic concept underpinning the new policy orientation was using market mechanisms to develop a more productive economy. Market functions were employed to drive the allocation of economic resources under state macro control. The government was strong and effective, and the Communist Party of China (CPC) was in position to oversee and guide the course of development. The production of consumer goods was expanded, as well as the provision of social services. Market mechanisms expanded production, enhanced technologies, and promoted overall efficiency and productivity, but the party and state maintained the socialist legal system and public ownership of both land and core sectors of the economy. This hybrid approach was called "socialism with Chinese characteristics."[80]

It was recognized, however, that the use of markets, even for the positive goals of economic development and supervised by the party and state, could generate serious challenges and potential contradictions among the people. Deng Xiaoping famously said, "to get rich is glorious," but, "some will get rich before others." He understood that, for a time, the new policies would create inequalities of income and wealth within the population, even as material standards of living improved for everyone. Perhaps less understood at the outset were problems of environmental stress and social tension that might be unleashed in rapid growth. These contradictions developed through the 1980s, precipitating a crisis in 1989. Both domestic and foreign forces hostile to socialism sought to exploit and manipulate the crisis to overthrow the government of the People's Republic and launch a "color revolution" to bring to power political elements more amenable to global, especially American, capitalist interests. The government and the party weathered this storm, and reform was reenergized in the 1990s, evolving and adapting to changing circumstances.

New Undertakings in the 1980s

At the beginning of the first decade of the reform era, new policy initiatives were launched in a range of areas. The agricultural sector was reconfigured, state-owned enterprises were restructured, special economic zones were established to experiment with direct foreign investment, and a small but growing sphere of private enterprise was created. New policies on population management were put into place.

Before exploring each of these developments, however, the question of political orientation needed to be considered. Deng Xiaoping's leadership and the policies he promoted represented a sharp break with the style and substance of political affairs during the first three decades of the People's Republic of China (PRC) under Mao's leadership. In 1979 the CPC set up an internal working group to assess the legacy of Mao's leadership; and to forge a new consensus within the party and in society at large in support of the dramatic changes about to be undertaken. In July 1981 the party published its "Resolution on Certain Questions in the History of Our Party Since the Founding of the People's Republic of China." This remains the definitive official position regarding Mao's legacy and his contributions to the revolution and the development of the country. The resolution was critical of certain decisions and actions by Mao, some of which were characterized as "grave errors," but it also asserted that his contributions to the revolution far outweighed his mistakes. It essentially argued that Mao played a positive role until the late 1950s, but that the Great Leap Forward and the Cultural Revolution had been misguided and caused great damage to the socialist cause. It is often said that the resolution judged that Mao had been 70 percent right and 30 percent wrong, though no such precise formulation appears in the document.[81]

As this review of party history and Mao's legacy was carried out, the new era of "reform and opening" was getting underway. Policy experiments in agricultural management were undertaken in Sichuan, Deng's home province, as early as 1978, and by 1982 were expanded to a national basis. The state retained ownership rights over all land, but decisions about what to grow and how to organize production were largely devolved to the household level. This came to be known as the "household responsibility system." Townships or counties would no longer function as collective units of production or accounting, and would enter into contracts with individual households. Certain basic commodities, primarily grain, would be provided for state purchase at set prices, but production above rather modest quotas would be available to be sold in newly developing markets. These changes yielded quick rises in aggregate yields and generated higher incomes for most rural households, though these gains plateaued after three or four years.

Over time these initiatives led to profound changes in the agricultural sector. Leasing rights could be transferred to individuals or even organizations outside the original household. This resulted in the aggregation of larger units of land in the hands of some families, and eventually to the development of an agribusiness sector that bought up leases to create economies of scale in production. As young people from the villages began to migrate to the cities in search of higher paying employment, many of them alienated their leases, though retaining their *hukou* household registrations. Household incomes and material standards of living rose steadily, though not as rapidly as in urban centers, after the initial surge in the early 1980s. As more and more people sought economic opportunities in the cities, the proportion of the population living in rural communities declined, falling to around 46 percent by 2020.

In tandem with the marketization of agricultural production, reforms were launched in the state-owned enterprises (SOE). These comprised the vast majority of manufacturing operations and functioned as comprehensive units of social provision and industrial production. Large-scale enterprises especially incorporated job security, residential space, health care, education, retirement services, and consumer outlets within their operations, in what was often referred to as the "iron rice bowl." The reform program now sought to subject these enterprises to the "discipline" of the market, to make them more efficient and productive. Management was restructured and evaluated based on the profitability of the business. Social services were spun off, largely to newly developing private providers in the case of medical care, or reorganized public educational institutions. However, these, too, were subjected to new forms of management in the interest of reducing operational costs. Eventually, even housing was transformed, with apartments provided by employers being sold to their occupants at low prices. These apartments then became available for resale in an emerging property market. These reforms weren't implemented overnight, and the evolving new policies and practices were often provisional and experimental, sometimes characterized as "crossing the river by feeling the rocks" *(mozhe shitou guohe)*.

The process of reform in the industrial sector generated significant social contradictions. Beyond stripping away social services formerly provided as part of workers' employment packages, the new management regime also discharged millions of workers as enterprises sought to maximize individual worker productivity and enhance their bottom-line profits. Especially in the northeast—the most heavily industrialized part of the country—there was widespread discontent. The state provided retraining and other services to aid discharged workers seeking new employment or starting small businesses, but the pain of this transition was still often intense. Many older workers simply settled into a kind of early retirement, with reduced incomes and straightened material circumstances. Others joined the internal migration that developed as people moved to new factories in the special economic zones (SEZ) along the southeast coast.[82]

One of the main objectives of the reform program was to gain access to foreign production technologies and organizational models. The SEZs were initially conceived to allow some direct foreign investment in China within strictly limited geographic spaces, insulated from the rest of the domestic economy. The first two were established in Shenzhen, northeast of Guangzhou, and Zhuhai, to the southwest of that city. Guangzhou had long been the site of the annual foreign trade exhibition, and was adjacent to Hong Kong, then still a British colony and seen as a likely gateway for investment from other Asian countries. Foreign corporations were invited to open joint-venture production units in the SEZs, with their output to be exported to global markets. Foreign businesses were allowed to repatriate profits, but obligated to share information about their technical operations and business practices with their Chinese partners.

The SEZs proved to be very attractive to foreign capital. Initially, investors came mostly from other Asian countries, especially Japan, Singapore, and corporations operating in Hong Kong and Taiwan. China began to acquire state-of-the-art technologies and accumulating rapidly increasing amounts of foreign currencies as export volume quickly expanded. The SEZs also became destinations for growing numbers of workers from villages across southern and southwest China. Many were young women, who sought employment in factories where they could earn wages considerably higher than income levels in their hometowns. As the SEZs flourished during the first decade of reform, the program was expanded to other locations, including the western suburbs of Shanghai.

The 1980s were characterized by reform in the agricultural sector, the reconfiguration of SOEs, and the development of the SEZs, as well as the initial stages of privatization of some social services, and the conversion of others from the provision by work units to generic public services like schools and hospitals. These policies yielded results in rapid rises in economic output, with gross domestic product increasing between 7 and 13 percent annually. In addition to measures aimed at economic growth, China adopted policies to reduce the rate of population increase that consumed much of the growth achieved during the first thirty years of the PRC. Referred to as the "one-child policy," it was a complex set of regulations and guidelines to encourage families to have fewer children, to maximize the gains in material standards of living that the reform policies were anticipated to generate. Couples with urban household registrations were allowed to have one child, with exceptions such as for a child born with serious medical issues, or of course, the birth of twins or triplets. Couples with rural household registrations could have two children, though if the first child was a boy, they were urged to consider having only one. Members of China's fifty-five ethnic minority communities were not subject to these limitations. As a result, their proportion of the overall population grew from around 4.5 percent in 1980 to nearly 7 percent by the end of the program in 2013.

The one-child policy was dramatically successful in reducing the rate of population growth and allowing the benefits of economic development to drive strong improvements in the material standards of living for the people. Problems of implementation and abuses by local officials took place in a few instances, but the overall success of this program cannot be denied.

Contradictions of Reform in the 1980s

As China's economic development accelerated through the 1980s, various policies and practices evolved that generated serious contradictions within social and political life. Some sectors of society saw improvements in incomes and living conditions, while others were left behind, or saw their livelihoods deteriorate. Millions of workers in older SOEs, especially in the northeastern provinces, were laid off or pushed into early retirement. Vocational retraining programs aided some workers in finding new employment, while others migrated to the southern coastal provinces in search of work in the developing SEZs, but large numbers found themselves in dire economic circumstances. Many workers in jobs requiring higher education, such as doctors, legal services, journalists, researchers in science or technology institutes, teachers in public schools or universities, or workers in government administration, saw their incomes stagnate while the new joint-venture companies and private start-ups raised employee compensation.

Corruption by state and party officials emerged as a significant concern. China did not experience anything like the selling off of state and public assets that devastated the former Soviet economy in the 1990s, but the move to market mechanisms created opportunities for bureaucrats to exploit their power, to enrich themselves through bribes, gifts, or the exchange of favors. Deng famously said, "getting rich was glorious," but, "some people would get rich before others." Most felt this was acceptable if the difference was due to someone working harder or being more creative or innovative, but not if it was a matter of taking advantage of their official positions to extort money or favors from those seeking licenses, permits, or other kinds of government authorization for their activities. The privatization of services, leasing out land for development, or granting permits for private economic activities, were all occasions for making informal deals and arrangements to facilitate processes that should have been carried out properly in the spirit of "serving the people." The rapid pace of change and the often vague or underdeveloped legal and procedural infrastructure allowed widespread abuses of power to flourish. Much of this was tolerated as the "cost of doing business" by other officials higher in the system. Ordinary people who experienced these corrupt practices grew increasingly resentful and cynical about the path of China's development.

The emphasis on rapid economic growth and the desire to maximize profits and achieve immediate results also led to quality control problems, and emerging environmental stresses of air, water, and soil pollution. Profit maximization, in both SOEs and private or joint-venture enterprises, sometimes fostered cutting corners in materials, ignoring safety policies, adulterating ingredients, and other shortcuts in manufacturing and construction. Consumers found themselves needing to be wary of shoddy products, a problem that reached the level of building collapses, or major failures in food safety, or other family or personal goods.

As industrial production expanded and households increased demands for electricity and heating, the burning of coal accelerated, leading to rising levels of air pollution in the cities. Building new factories and using greater amounts of chemical fertilizers in agriculture poured more and more pollutants into rivers and streams, which seeped into underground aquifers and spread through the water table. Precipitation brought toxins from factory smokestacks and coal-fired power stations into the soil of farmland across the country. Environmental stresses began in the 1980s, but were sufficient to add to the emerging sense of frustration about the effects of the rapid development of the economy through the reform program.[83]

The first decade of the reform period saw continuing debate within the national leadership. Deng functioned as the final arbiter of disputes and was in overall control of policy decision-making. But there was a range of opinions within the party center, with figures such as Hu Yaobang and Zhao Ziyang advocating greater opening to global market forces and a general deregulation of the domestic economy.

Others, including Li Peng, Hu Qiaomu, and other senior comrades, were concerned about the pace of reform and the negative social and political impacts of marketization. A central question in these debates was price setting, largely released to market forces. This generated rapid inflation by 1987-88, especially in the cities.[84]

By the late 1980s political tensions from the effects of the reforms gave rise to demonstrations on university campuses in various parts of the country. Many students and their professors shared feelings of being marginalized or left behind in the surge of economic transformation. This, combined with broader public frustration about corruption and the abuse of power by some state and party bureaucrats, came together in the spring of 1989. In a crisis begun as another round of student protests, it escalated into a challenge to the existence of the People's Republic. This was manipulated by domestic advocates of all-out conversion to capitalism as well as foreign, primarily American, overt and covert efforts to overthrow the government of China and destroy the Communist Party.

The 1989 Protests

On April 15, 1989, Hu Yaobang collapsed and died of a heart attack during a meeting of the Political Bureau of the CPC. Over the previous two years, he had become the focus of contention over the pace and direction of the reform efforts. He advocated for faster and more thorough reform, meaning a greater degree of deregulation and further opening the country to direct foreign investment. He was also outspoken concerning the anxieties expressed by students about their role in China's future. This brought him into conflict with other leaders in the Party and the government. He was removed from his most senior position, but remained a member of the Political Bureau, the party's highest policy-making body.

In the days immediately following Hu's death, students from universities of the Haidian district in northwest Beijing, including Beijing and Qinghua Universities, the most prestigious in the country, held rallies to express their admiration for him. They marched to the city center, Tiananmen Square, to place flowers and placards at the Monument to the People's Heroes, the focal point of political gestures at different times in the history of the People's Republic. They also used these occasions to call for further economic and political reform.[85]

Toward the end of April, the *People's Daily* ran an editorial characterizing these demonstrations as unpatriotic. This infuriated many of the students, and their protests accelerated. Over the next few weeks, several thousand marched to Tiananmen Square and established a tent city. This disrupted the regular flow of traffic through the central city and the normal functioning of government institutions nearby. As time went by, the students' demands grew more confrontational. The government initially tried to ignore them in the hope that things would simply calm down and return to normal. But as the weeks went by, government and party leaders sought a way to end the occupation of the square and defuse the emerging conflict.

The situation deteriorated in the middle of May. Soviet leader Mikhail Gorbachev was scheduled to pay a state visit, and Chinese leaders were eager to show off the successes of the reform era. But when Gorbachev arrived, he could not be given the grand reception planned for him at the Great Hall of the People, west of the square. He was brought into the building by a side door. This was a great humiliation for China, and the leaders of the government and the party. The international press in Beijing, present in large numbers to cover Gorbachev's visit, devoted their efforts to glorifying the students and ridiculing the Chinese leaders. American journalist Dan Rather became the number one cheerleader for the protests, portraying them as the prelude to the fall of the Communist Party and the overthrow of the government.

State and party leaders were somewhat divided over how to end the occupation of the square. Zhao Ziyang, Hu Yaobang's closest ally, pushed to accommodate the students' demands, but most of the leadership increasingly saw the protests as aimed at overthrowing the existing system and replacing it with a Western-style liberal capitalist one. Statements by the most extreme student leaders called for bloodshed to provoke an anti-government revolution. As US and other Western journalists gleefully pushed the idea of imminent collapse of the PRC, the students increasingly used English for signs and posters they held up for the foreign media cameras. At the end of May, a large statue meant to look like the American Statue of Liberty was rolled into the square, to the delight of Dan Rather and the other media pushing the agenda of regime change. The extent of clandestine support for the protests by agencies like the US National Endowment for Democracy (NED) or other behind-the-scenes foreign manipulation of the situation is unknown, but the protesters' methods were largely consistent with those developed and deployed by the NED in the many "color revolutions" they've sponsored around the world.

A final effort to broker a resolution of the crisis took place at the Great Hall of the People when government and party leaders met with protest representatives. But student leaders, at least one of whom was wearing pajamas, weren't willing to compromise. The intransigence of the protest leaders, who in other statements expressed their hope for violence to advance the overthrow of the government, drove state and party leaders to declare martial law and authorize the People's Liberation Army (PLA) to clear the protesters from the square. Much of the city's transport system and economic affairs had been paralyzed for a month and a half, and there seemed to be no way out of the deadlock in Tiananmen Square.

On the night of June 4, PLA units moved from the suburbs into the center of Beijing. On the west side, there was extensive fighting along Chang'an Avenue, the main east-west route across the city. Army vehicles were attacked with firebombs, and trucks with soldiers were stopped and attacked. Weapons were seized from troops who had been injured or killed and then turned on other PLA troops. Soldiers returned fire and pressed on toward the square. They reached the Great Hall of the People around 2:30 a.m., and the square was soon surrounded. Only a few people remained, and they were offered the chance to depart by a gap left for them at the southeast corner. Most chose to do so, though a handful remained, who were then arrested.

The most credible reports of these events, drawn from both participants and from hospital medical personnel who treated the injured protesters and soldiers, suggest that between six hundred and eight hundred people were killed during the fighting, of whom at least two hundred were soldiers. No one, however, was actually killed in Tiananmen Square. Western media was filled with fabricated accounts of tanks running over people in sleeping bags in the square aimed for maximum emotional impact. Even at the time, several Western journalists acknowledged that the sensational claims about events in the square were not true. The famous photo of a man carrying shopping bags standing in front of a column of tanks, often thought to represent a confrontation during the June 4 events, was actually taken a few days later, when the tanks were withdrawing from the city. The man

in the photo subsequently moved on and was not subject to any reper-
cussions. There is no question there was intense fighting in Beijing on
the night of June 3-4, but it was not a massacre of peaceful protesters
by the army.

The political events of April-June 1989 were an expression of the
contradictions arising from the economic reform and the complex
shifting relations between various sectors of the working class, as well
as divisions of opinion and policy perspectives within the Communist
Party. There were forces opposed to the party's leading role who hoped
to overthrow the government of the People's Republic. They were
aided and encouraged by American and other outside actors, overtly
and covertly, in their efforts to manipulate the ideas and actions of
young people to create a crisis for the party and the government. The
occupation of Tiananmen Square, at the heart of the capital of the
People's Republic, disrupted the normal economic and social life of
the city and the regular functions of government for weeks. When the
movement was taken over by the most radical and anti-government
leaders toward the end of May, it was clear they would accept no compromise. Using the PLA to clear the streets and restore normal life to
the city was not something anyone would have wished for, but was
viewed as necessary to end the crisis.

Reaffirming Reform

In the wake of the suppression of the counterrevolutionary crisis,
China faced two immediate challenges. One was to resolve the contra-
dictions within the party leadership and determine the path forward.
The other was the international reaction to the events of that spring.
Once again, Deng was the key figure in shaping these decisions and
political processes.

At the central leadership level, there were several major changes.
Zhao Ziyang and a few other leaders—who were seen as having
encouraged the divisive forces within society and pushing for policies
that would undermine the party and the government in preserving
the system's socialist nature—were transferred from their positions.
Jiang Zemin, the mayor of Shanghai, managed the spring protests
there to prevent the legitimate concerns of people about the effects of
the reform policies from developing into antagonistic contradictions.
He replaced Zhao as general secretary of the CPC. The party under-

took an internal process of criticism and self-criticism to analyze and understand the experience of the previous period, and to determine the best course to follow in the coming months and years. This process continued through the rest of 1989 and into 1990.

American imperialism and its allies were frustrated with the failure to overthrow the PRC and usher in a new and weaker China, to be effectively subsumed within the global capitalist order. The mass media, eager to foment a reactionary revolution, now pushed a narrative of condemnation, seeking to isolate and punish China. Trade with China declined, though it continued to be an important destination for investment and source of exports for Western markets. Tourism and educational exchanges were largely shut down through the summer and fall of 1989, though both revived by early 1990. American capitalism faced its own contradictions, and still hoped that in the long term China would be brought into conformity with the bourgeois world order. Media criticism and diplomatic isolation was designed both to sanction China and as a show of moral self-righteousness to convince Western audiences of the superior virtues of the "liberal democratic" regime. But the prospect of profits from commodity production for export to US and European markets and the sale of Western goods to Chinese consumers was too powerful to be set aside for long. After a season of political posturing by political and business elites, trade and tourism resumed and grew beyond their pre-1989 levels.

From January 18 to February 21, 1992, Deng Xiaoping undertook an inspection tour of the SEZs in southern China. In the course of this tour he made several speeches articulating China's continuing commitment to the policies of reform and opening to the outside. This was a public demonstration that the internal process of policy debate within the Party had concluded, and it was time to return to developing the economy through market mechanisms and engagement with global capitalism as a source of technology and investment. The recommitment to reform policies demonstrated by Deng's 1992 "Southern Tour" remained the orientation of the Party and the government for the following three decades, at times with elaborations and revisions, but consistent in its fundamental conception. Deng continued as the guiding spirit until his death in 1997.

In the following final chapter, we'll consider the state of affairs in contemporary China as it's developed through the reform era,

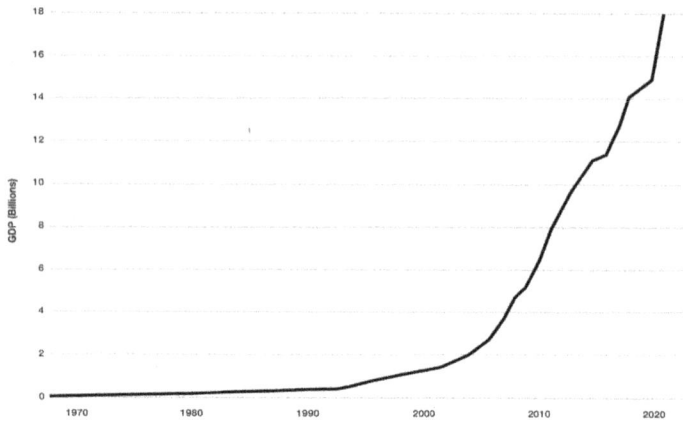

The PRC's GDP rose dramatically after 1992. (Chart: Hannah Craig)

China's relationship to the rest of the world through its continuing engagement with the global capitalist economy, and its efforts to foster connection and interaction with the "developing" nations of the world through initiatives like the Belt and Road Initiative, especially since 2012, under the leadership of Xi Jinping.

CHAPTER FIVE

CHINA IN THE TWENTY-FIRST CENTURY

Biding Time, Building Capability

When Deng Xiaoping died on February 19, 1997, China was firmly set on the course of reform and opening he had championed since 1979. Significant gains were made in developing the economy and raising the material standards of living for the people. There were also serious contradictions and challenges that arose over the reform era's first two decades. China's entanglement with the global capitalist system and often its strained relationship with the United States—the hegemonic imperialist power—yielded major benefits, but also put China in a subordinate position. Indeed, Deng recognized and accepted it as a temporary but necessary situation. He told Jiang Zemin and other leaders that China needed to "bide its time and build its capabilities," keeping a low profile internationally and not openly challenging America's predominant role in global affairs. Over the next fifteen years, Jiang and his successor Hu Jintao adhered to this precept, emphasized China's "peaceful rise" and pursued further integration into capital's established order by joining the World Trade Organization and other economic and political bodies.[86]

These years saw China's economy grow remarkably, with the GDP expanding over 10 percent a year on average, though there was significant variation year to year. Problems associated with rapid growth, as discussed previously, persisted, but as China accumulated wealth, the People's Republic of China (PRC) government and the Communist

Party of China (CPC) were able to address them. Expanding national wealth enabled China to enhance its military capabilities, upgrade weapons systems, and improve training and conditions of service for the People's Liberation Army (PLA), while its military budget remained a fraction of that of the US.

As the economy boomed and military capacity strengthened, Western observers questioned their assumptions about the future of the country's political system. American politicians, academics, and media pundits long proclaimed their expectation that economic reform and engagement with the global capitalist system would lead to political "liberalization" in China. They assumed China would undergo a fundamental transformation of its system, the kind of "color revolution" that took place in other countries, opening them to capitalist enterprises and finance capital's unrestrained activities. But as the years went by there were no signs of moving in such a direction. Some policy analysts and media commentators wondered if China's "rise" was such a good thing: Was it on the way to becoming a serious rival to American hegemony?

The first decade of the twenty-first century encompassed significant developments that seemed contradictory. In 2002 the CPC constitution was amended to allow capitalists to become party members. With China's accession to membership in the World Trade Organization and the ongoing process of reform and opening, deepening the country's entwinement with the global capitalist system, many observers saw the party's admission of capitalists as a sign of a movement toward both economic and political "liberalization." However, it can be better seen as a move to bring bourgeois elements into the democratic-centralist discipline of party life. Capitalists remain a small portion of overall CPC membership, and having them within the party means they are subject to oversight and guidance, with an obligation to adhere to government and party policies.

When the 2008 financial crisis struck the US and Western economies, China was mostly insulated from the financial debacles that devastated banks and investment houses, which wiped out massive quantities of spurious financial assets, and caused suffering for tens of millions of workers. China maintained control over its financial sector, meaning the effects of the collapse of the derivatives-driven bubble were minimal. The economy's socialist core and the party's guiding role in economic affairs protected much of the domestic system from the severe downturn plaguing the capitalist heartlands.[87]

Nonetheless, the dramatic drop in demand for goods produced in China and exported to American and European markets meant that some twenty million workers in China were laid off from their jobs. This was a major social challenge for the country. In a capitalist economy these workers would have largely been left to fend for themselves. China's socialist infrastructure includes a system of household registration that means every citizen has someplace to go for basic social needs, such as housing, education, or health care. The workers laid off in 2008 were largely drawn from rural areas, part of the so-called floating population, and their household registrations remained in the villages from which they came rather than the cities where they worked. Laid-off workers could return to their villages and have a place to live, schools for their children, and access to health care. The level of provision in rural areas remains lower than in urban areas, but these millions of people made it through this difficult period without being reduced to homelessness or poverty. As the economy revived from 2009 onward, they were able to return to their previous employment or seek new opportunities.

After thirty years of reform and opening policies, China made great progress toward its goals of developing its productive economy and raising the material standard of living for its people. This progress came with significant costs, including corruption, environmental stress, increasing inequality, and other social issues. The People's Republic drew in huge amounts of foreign investment and acquired a great deal of state-of-the-art production technology. But this also meant it had to accept a subordinate role in relation to the global capitalist order. As it entered the second decade of the twenty-first century, China became more prosperous, though still a developing country with per capita incomes far below those of the US or Europe. It also became more self-confident. Economic planning shifted toward emphasizing greater domestic aggregate demand, rather than relying primarily on production for export to drive growth. China became less dependent on foreign markets and less deferential to American power and interests. At the same time, US political analysts became increasingly wary of China's growing power, and doubted their fantasies of an imminent transformation of China into a free-market enterprise

AMERICAN PIVOT TO ASIA

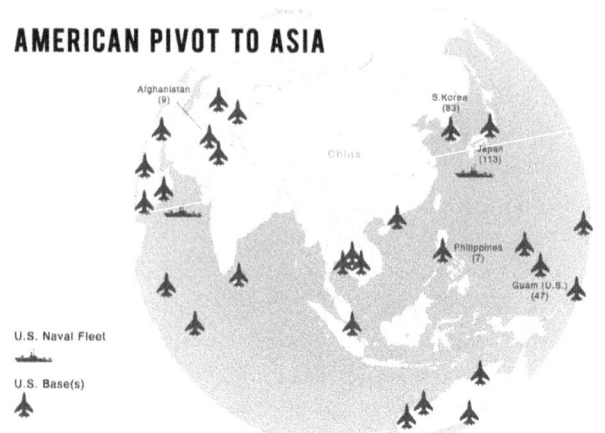

America's "Pivot to Asia" containment strategy (Map: Tina Duong)

zone with a post-CPC government. The stage was set for major shifts in the global order.

Turning Points to a New Era

The years 2011 and 2012 marked the beginning of a new period in China's modern history and its relationship with the United States. In November 2012, Xi Jinping became general secretary of the Communist Party of China and was named chairman of the Central Military Commission. The next year he was also elected president of the PRC. He was seen as the logical successor to Hu Jintao since about 2008, but his formal assumption of these leading positions made him the primary figure in the hierarchy of the party and the government. He quickly began to promote a series of policies and pursue political initiatives that led to a new era of confidence and assertiveness for China and the CPC.

Even before Xi's elevation to the leading positions, US President Barack Obama and Secretary of State Hilary Clinton announced a significant shift in Washington's attitude. In November 2011 Obama unveiled his "Pivot to Asia" plan, which involved repositioning US military forces, a shift away from the established focus on Europe, and the more recent priority of the "War on Terror," to a deployment to contain and constrain China. That same month *Foreign Policy*, a mouthpiece of the foreign policy establishment, published an article by Clinton, in which she proclaimed the twenty-first century to be America's "Pacific Century," signaling a new emphasis on projecting power in the Pacific region and containing the rise of China, seen increasingly as a challenge and a threat to US global hegemony.[88] This new hostile posture was expanded and increased in the following decade. Fear and hatred directed at the PRC and CPC is a bipartisan feature of American political culture, reflecting the ruling-class realization that China won't roll over and allow domination by global capital. It will pursue a socialist future and may become a model for developing countries around the world that hope to avoid being completely subsumed under the American-led capitalist imperial order.

China in the Current Era

When Xi Jinping assumed the leading roles in the party and government, he faced the challenge of corruption by state and party officials. The problem was not only the abuse of power by those in positions of public responsibility undermining the prestige and legitimacy of the socialist market system. It also diverted resources from the critical tasks of developing the economy and improving the livelihoods of the people. The 2012 anti-corruption campaign continues to the present, with dramatic results. Officials at all levels, from the local to the highest central organizations, have been investigated, arrested, tried, and convicted of serious crimes and misconduct. New policies and regulations have placed restraints on the conduct of official business and reduced opportunities for bribery and influence peddling. Corruption hasn't been eliminated, but it has been reduced, and it's clear that officials who engage in corrupt behavior do so at great risk to their careers, reputations, and livelihoods. These measures have been welcomed by ordinary people, who understand corruption is a serious problem and respect the efforts of the party and the government to address it.[89]

Over the last decade, China undertook major programs to deal with many issues that arose in the course of reform and opening, corruption being only one of them. Massive resources were devoted to environmental stresses, cleaning up the air and water to reduce damage to public health from pollution. This has been tied to efforts to develop and promote the use of alternative energy. China has massive reserves of coal, and its use as an energy source remains widespread. But the country has increased the production of alternative energy, causing coal production

Wind farm in Xinjiang (Photo: Zhao Ge, Xinhua)

to fall steadily in the energy economy. Wind power, solar and hydroelectric energy, and nuclear power production have all been important.

China is now the leading producer and consumer of wind and solar energy. It is the top producer of solar power panels for the global market, though the US places prohibitive tariffs on them. This harms American efforts to deal with climate change and global warming. China is also the leader in electric vehicle development, building infrastructure (including charging stations) along its highways to ensure adoption of this technology by consumers.

China made public commitments to reach its maximum carbon output no later than 2030, and to reduce emissions to achieve carbon neutrality by 2060. News reports document efforts to meet this goal, with specific plans for industries and other enterprises. The efforts must continue, and they will entail a complex mix of uplifting the material standards of living with reducing energy and resource consumption. This could play a critical role in solving planet-wide environmental problems, by providing a model and inspiration for people in other developing and advanced countries.[90]

In the last decade, China embarked on a new program of global economic cooperation and development aimed at the poorer countries of Asia, Africa, and Latin America. These countries face the after-effects of imperialism and colonial rule or the negative impacts of the

Countries of the Belt and Road Initiative (BRI)
Note: Up to March 2022

The Belt and Road Initiative (Map: Tina Duong)

structural reform policies of the International Monetary Fund and the World Bank. These institutions routinely impose austerity measures as conditions for loans or investments that gut social programs and produce massive suffering for working and poor people.

In 2013, Xi Jinping announced the inauguration of the Belt and Road Initiative (BRI). This is a massive, long-term program building new infrastructure for trade and economic development through loans and investments. China works with countries in Asia, Africa, Latin America, and Europe, building ports, railways, airports, hospitals, schools, and even sports facilities. Developing infrastructure and production capacity will benefit both the host countries and China. BRI is not a charitable activity, nor a welfare program.

China hopes to benefit by growing its domestic economy and expanding trade relations. It is a mutual benefit for all the participants. Importantly, it's outside the long-established nexus of American-centered global capitalism, providing alternative options for countries that want to improve their people's livelihoods without becoming subordinate to American interests.[91]

The struggle against corruption, the efforts to tackle environmental issues and develop alternative energy, and the Belt and Road Initiative are three main features of the current era. All three form a part of the broader goal of improving the lives of the Chinese people. In 2020, a major milestone on the path toward this objective was reached when China announced the elimination of the last pockets of absolute poverty as defined by the United Nations. Over more than seventy years of the PRC's history, the people's material standard of living has steadily improved. In the period of reform and opening, more than 850 million people have been lifted out of absolute poverty due to economic growth. Health care, education, and housing have all improved and expanded. China remains a developing country, with differences in income and material conditions between urban and rural society, and inequality between the richest and poorer citizens. But the accomplishments of the PRC and the CPC are unequaled in any major economy or country, and prospects for the future, though challenging, look bright. Why, then, do Americans hear only negative things about China? Why is it relentlessly portrayed as a dictatorial state, bent on the oppression of its people and the expansion of its power over the world?

Toward a New Cold War

As China became more self-confident and assertive in its relations with the wider world, moving beyond the paradigm of "biding time and building capabilities," ruling-class politicians and the bourgeois media steadily hardened their hostility. The Obama administration pushed ahead with Pivot to Asia military redeployments and initiatives like the Trans-Pacific Partnership (TPP). The TPP created multinational free-trade agreements among Asian states, explicitly excluding China in an effort to stall its economic development. When Donald Trump became US president in 2017, he launched a trade war with China, though the regressive tariffs imposed did more to hurt American consumers than the Chinese economy. He unleashed a torrent of anti-Chinese racism, repeatedly slandering China for the outbreak of the Coronavirus pandemic. With no evidence, he even accused China of making the virus in a lab and releasing (or allowing) it to escape to the outside world. The bogus "lab leak" myth has been continued by the neoliberal Biden regime. There were incidents of American diplomatic figures publicly lecturing the Chinese foreign minister Wang Yi and his colleagues that China should conform to a "rules-based international order," with rules of course, set by American imperialism. Both parties in Congress engaged in anti-China posturing. Resolutions from both sides of the aisle condemned China for a wide range of alleged misbehavior and imposed economic sanctions. These actions became part of a New Cold War on China. Biden's secretary of state, Antony Blinken, explicitly stated the era of engagement with China was over. The American political elites declared it an adversary and continue their demonization campaign with a litany of accusations. The American people, as workers, have more in common with the Chinese people than the capitalists who enrich themselves by expropriating the wealth US workers produce.

But, they are subjected to a relentless stream of anti-China propaganda designed to foster fear and hatred to prepare for possible conflict, to drive a wedge of mistrust between workers on opposite sides of the Pacific. There are many variants of these accusations, with new charges invented and propagated regularly, so, it's useful to address some of the issues which are repeatedly invoked.

China is said to both be oppressive of its people at home and aggressively expansionist abroad. Both accusations are a matter of American elites projecting their behavior and attitudes onto China more than anything taking place in the real world. Portrayals of an authoritarian and neo-imperialist China by politicians and pundits rely heavily on Americans not knowing much about it. Few Americans read or understand the language, and have no understanding of the intellectual and political culture of the country. In China, bookstores and newsstands are filled with a wide and dynamic range of materials and information, including diverse writings, both fiction and nonfiction, by local authors, and translations of works by foreigners. Chinese television is lively, and hundreds of stations produce programming, often in local dialects. Many shows, especially popular dramatic series, deal with contemporary social issues. The internet is an energetic arena of popular expression, through which "netizens" carry on debates and discussions, and sometimes organize protests and political campaigns about their concerns and aspirations. There certainly is some media oversight, to prevent the spread of false information as well as pornography and such. This is not to say China's public discourse—print and electronic media—is exactly like that in the US. But given the degradation of public discourse, politicians' and special interest groups' lies and manipulations, surging tides of fake news, and corporations' generation and manipulation of "consumer needs" pursuing ever greater profits, perhaps this is not a bad thing.

It is often said that China has no independent or real trade unions or other workers' organizations. This argument assumes its socialist economic and legal system is somehow illegitimate. Since the 1950s, the All-China Federation of Trade Unions (ACFTU) has been the umbrella organization representing workers' particular needs and interests. Because the government, working with the Communist Party, represents the political power of the working class, the ACFTU has official status. It is a legally established body able to articulate the workers' interests in daily government operations and public discourse. It produces the *Workers Daily* and other publications. It carries on industrial policy oversight, occupational health and safety issues, the collection and dissemination of statistical information, and many other tasks.

Because the People's Republic is formally a socialist state, the ACFTU is an integral part of the system. Those who say Chinese workers have no real voice, even some on the left, clearly fail to understand (or do not accept) how the established power of the working class functions. There's a long road ahead to reach a fully developed socialist economy and society, and as noted above, there are challenges and contradictions involved in that trajectory. But to argue that only a union or other grouping, outside the legally established mechanisms of representation, would be legitimate is to reject core precepts of building socialism, of the realities of Chinese political life.

Ethnic and Religious Issues

A common accusation against China's domestic situation recently deals with religion or ethnicity, mostly concerning Xinjiang or Tibet. Again, China is said to oppress people based on religious beliefs or ethnic identity. This is neither the theory nor practice of the government nor the CPC. There are fifty-six recognized nationalities, all of whom are full citizens of the People's Republic. The vast majority are Han, about 94 percent of the population of 1.4 billion. The other fifty-five ethnic groups altogether account for 6 percent, with a few, including the Zhuang, Mongols, Tibetans, or Uighurs, constituting the majority of these, each with between five and ten million people. Some ethnic groups, such as the Ewinki, Uzbeks, or the Thai-speaking communities in the southwest, have only tens of thousands of members. Ethnic policy for minority peoples is a complex blend of respecting the traditional cultures of each group while providing opportunities for participation in the overall development of China's modern economy. For groups with distinct languages, education is in both the minority language and standard modern Chinese—called Mandarin. Balancing the preservation of traditions with the process of modernization has been challenging. Still, it has largely worked well, certainly compared to the marginalization and impoverishment of indigenous peoples in more developed capitalist countries like the United States and Australia.

In Xinjiang and Tibet, particular circumstances have generated special tensions. There are similarities, but also important differences. Tibet formed part of various Chinese states since the seventh century. During imperial times there was a close relationship between religious authorities in the theocratic Tibetan system and the emperors of different dynasties. Tibet has been a part of the territorial state since the seventeenth century, a fact internationally recognized under the Qing dynasty, the Republic of China, and the People's Republic.

Xinjiang has a more complicated history. The geography now called Xinjiang was incorporated into China during the Han (202 BCE-220 CE) and Tang (618-907 CE) dynasties. At other times, it was ruled by different ethnic and religious groups, including Turkic peoples, Mongols, and Sogdians from Central Asia. In the eighteenth century, the Manchu-ruled Qing dynasty conquered the region and destroyed the Buddhist Zhungar state, then occupied much of the territory. It was then integrated into its multi-ethnic polity. The space vacated by Zhungars was then taken over by Muslim groups, including Uighurs, Kazakhs, Kirgiz, and Tajiks.

When the last dynasty was overthrown in 1911-12, and as the abortive Republic faltered through 1912-49, central authority in China was weakened. Both Tibet and Xinjiang were administered by local authorities still loosely connected to the governments in Beijing or Nanjing. In Tibet, the power of Buddhist monasteries controlling the land and extracting wealth from the people's labor continued to dominate under the Dalai Lama's leadership. In Xinjiang, a Chinese warlord exercised power, but in the 1930s there were short-lived rebellions by Islamic forces wanting a separate religious state called the

Eastern Turkestan Republic. The Soviet Union also developed influence there during these years.

The establishment of the People's Republic in 1949 involved the formal succession of the new government to the territories of both the Qing dynasty and the Republic of China. The PRC moved carefully to extend administrative control over Tibet and Xinjiang, designating them autonomous regions under the policy of granting special status to areas with minority groups a majority of the local population. Since, national policies balancing the preservation of traditional language and culture were implemented, as noted above. Both regions have experienced contradictions, sometimes causing serious social disturbances.

Tibet was a theocratic state, meaning the government and the religious hierarchy were the same. Buddhism was the state religion, and the monasteries were effective rulers of the country. In 1950, the central PRC government negotiated an agreement with the Dalai Lama and his advisors, giving the local administration in Tibet special exemptions from certain policies, including land reform. In 1959, though, a rebellion fomented by conservative monastic elements tried to separate Tibet from the rest of the country. The United States trained and infiltrated saboteurs and other secret agents into the region throughout the 1950s. When the rebellion was suppressed, the Dalai Lama fled to India, where he has maintained a "government in exile" ever since.

Since 1959, Tibet has seen extensive economic development, and people's lives have improved considerably. As in the rest of the country, there have been great gains in life expectancy and public health overall, as well as in education and the material conditions of daily life. As Tibet developed and more people, especially the young, have become engaged with the rest of the country and the wider world, traditional culture has increasingly been displaced by more modern ideas and behaviors. The monasteries face a declining societal role, as fewer young men want to become monks. This has led them to protest against modern development, though the modernization policies are quite popular with most Tibetans. Western bourgeois media portrays this as repression of religion, but it's actually the monasteries trying to reassert the power they used to wield over Tibetan society.[92]

Xinjiang's situation also has contradictions between the old religious authorities and the impacts of modernization, but is further complicated by the rise of militant fundamentalist Islamic movements

The regional political situation, with ongoing concerns about ter-rorism and Islamic separatism, intensified in the wake of the US defeat in Afghanistan, will likely remain a focus of government programs, and will no doubt be attacked by American politicians and journalists. The struggle to defeat the demonization of China will also carry on.[93]

Challenges in the Maritime Realm

Three important areas of concern are spread along China's south-east coast and the adjacent maritime zone. Hong Kong faces turmoil fomented by anti-communist forces, supported by US clandestine and semi-clandestine groups and organizations. Taiwan, occupied by the remnants and successors of the Nationalists since 1948, officially recognized as an integral part of China by the United States and the United Nations, has become a renewed point of friction in Washing-ton's New Cold War. The US has aggressively ramped up its presence in the South China Sea, a maritime territory claimed under Chinese sovereignty by both the Republic of China before 1949 (and continu-ing today) and the People's Republic. China has positions on islands across the South China Sea and plans economic development of under-sea resources. Since a large proportion of the country's oil supply is shipped through these waters, China is also interested in maintaining open navigation of the seas. These issues have been intensified by both Democratic and Republican administrations over the last decade as part of their efforts to contain and constrain China's development.

Hong Kong was a colonial possession of Great Britain for about 150 years. The island was taken by Britain during the Opium War of 1839-42, and additions to the colony were made, first in Kowloon and then with the New Territories, leased until 1997. The lease's expiration led to Hong Kong's return to Chinese sovereignty that year. While a colony, Hong Kong was run by a governor appointed by London. The local people had no voice in the affairs of their city, though rich busi-nessmen wielded influence through their interests in preserving the power of capital. There were no elections and no representative bodies.

When Hong Kong became a Special Administrative Region (SAR) of China in 1997, it was governed under the Basic Law. This gave Hong Kong its distinctive legal and political systems while rec-ognizing the territory as an integral part of the People's Republic. This is the famous "one country, two systems" policy that remains in place today. The Basic Law includes provisions to introduce elections for a Legislative Council on a step-by-step basis and to develop the political culture of the former colony incrementally. The specific organization of elections has been revised steadily, and the direct participation of the people has expanded over the last twenty-five years. There can be disagreements over the pace of these changes, but terms of the Basic Law remain in place and are the legal basis of Hong Kong's administration.

However, some elements within Hong Kong are hostile to the PRC's socialist system. They want to see Hong Kong as an independent state, with a free-market capitalist system through which they could enrich themselves and promote their own power. The United States, always eager to find any means to attack China, funnels support to these groups and individuals through the National Endowment for Democracy, clandestine channels like the CIA, and others. Ironically, many young people in Hong Kong are frustrated over economic issues like the high cost of housing and limited employment opportunities—the result of the SAR's particular property system and economic "openness." Legitimate concerns by some in Hong Kong have been manipulated by anti-China forces, backed by American funds with the advice and guidance of experts in fomenting "color revolutions" to open new areas for capitalist exploitation. Demonstrations have often been violent, with firebombs thrown in subway stations, blockades of the city's airport, attacks on police officers, and other destructive activity. Western media reports little of this destruction. There is no real prospect of Hong Kong becoming an independent country, and Washington knows it. The riots in the SAR don't serve the needs or interests of the local people, but provide a propaganda weapon for the US to further demonize China.[94]

Another arena for American interference and manipulation is the provocations carried out around Taiwan. The US government is legally committed to the reality that Taiwan is a part of China, the official position of both the Beijing government and the local authorities, though they disagree on who constitutes the formal governing entity. Over the seventy-two years since liberation, the relationship between the mainland and Taiwan has evolved, and in the last decades, there's been a great deal of interconnection between the two parts of the country. Hundreds of thousands of people from Taiwan now live and work on the mainland. Millions of tourists from both sides of the Straits travel back and forth. Billions of dollars of investment from Taiwan weaves ever-growing and -closer economic links with the rest of the PRC. Conflict between the central government and local authorities would be devastating for everyone concerned. Only US politicians, seeking any way to attack and thwart China's development, push for greater tension and possible confrontation across the Straits. Some political forces on the island are willing to play along with the Americans in hopes of advancing their own agendas, but this is a risky and dangerous game. Any military conflict over Taiwan would be massively destructive for the people of the island and would only serve US imperialist ambitions. Realistic scenarios for such a clash suggest China would prevail militarily, with significant losses for the US. Opposing US interference in the Taiwan-mainland relationship is fundamental to maintaining peace and prosperity for all people in the region.

American efforts to define the South China Sea as part of its global, imperialist domain and assert the US Navy as the world's maritime police force, seeking to provoke confrontations with China in the waters around the islands, constitute one more page in the anti-China playbook. Australian and British naval forces have also been recruited for this campaign. The United States has long defined itself as the core of the global maritime system, with the right and the capacity to control the high seas and project its power wherever it chooses. Several states assert claims to the islands and surrounding waters in the South China Sea, but the United States is not one of them, and has no legitimate interest in the area. Whatever the final determination of the claims by Vietnam, Malaysia, and the Philippines may be, the outcome should be a result of negotiations between the states in the region. American imperialist intervention and its plan to "contain" China must be resolutely opposed.[95]

The Road Ahead

In July 2021, the Communist Party of China celebrated the one hundredth anniversary of its founding. October 1, 2021, was the seventy-second anniversary of the establishment of the People's Republic. The country continues its development of a modern socialist economy, using market mechanisms to develop its productive capacities, with the CPC playing the guiding role. But the ultimate outcome of this venture remains to be seen. The party has taken significant actions to constrain the negative effects of contradictions arising from private capital as a part of the mixed economy. Regulatory oversight of enterprises like Jack Ma's Alibaba and the Ant Group, the suppression of for-profit educational services privileging wealthy families, and moves to control speculation in property and housing, demonstrate a commitment to a more equitable distribution of social wealth. China's clearly stated goals of limiting carbon output and developing alternative energy reflect a long-term drive to an ecologically sustainable future. The prioritization of public health shown by China's management of the COVID-19 virus, mobilizing state, party, and social resources to save millions of lives, makes the people-centered orientation of the government and CPC clear.[96] But the role of private capital remains significant, and the tensions and contradictions generated by the processes of capital accumulation may still threaten the success of China's socialist project.[97]

How can China's political-economic system best be characterized and understood? Many in the West, including some on the Left, see China having embraced global capitalism, as "state capitalist." This in many ways is just as problematic as the version promoted by US imperialism. The Chinese leadership refers to their system as "socialism with Chinese characteristics," a phrase with origins in Mao's revolutionary theory of the 1930s and the early years of the People's Republic. Xi Jinping regularly uses the phrase "socialist market economy" to describe China's project, as do scholars like John Ross and Yeo Yukyung, cited above. China's practical experiences in weathering global economic crises in 1997 and 2008, as well as its demonstrated commitment to prioritizing social welfare and public health over profits and economic interests in the Coronavirus pandemic, demonstrate the continuing resilience of the socialist infrastructure and legal system, as do recent efforts to restrain some private capitalist profit-seeking and accumulation. However one wishes to label it, China's efforts to build some version of a socialist future remain very much a work in progress, but one which is in progress.

A position of critical support, with the defense of China from US hostility and aggression as the main task, should be taken by comrades and friends. We recognize that there may be problems along the path to China's future, and that mistakes may be made along the way. We need to be prepared to acknowledge shortcomings, and extend fraternal critiques when appropriate. The process of building socialism is a long-term endeavor, as Marx, Lenin, and Mao all pointed out. In the present historical conjuncture, China represents one of the greatest potentials for actually reaching that goal.

PHASES OF CHINA'S SOCIALIST REVOLUTION

BY EUGENE PURYEAR

I n China the chasm between the goals of the revolution and their realization was, to put it mildly, yawning. The 1949 revolution took place in a country still containing elements of feudal relations. Most of the people were peasants, not workers. Landlords exploited millions who worked the land, despite land reforms carried out in territories liberated by the Red Army. The country was reeling from over one hundred years of being carved up by various foreign powers and warlords who ruthlessly exploited peasants and workers in the limited, export-oriented, hot-house industries on the coast.

This created overlapping imperatives for the Chinese Communist Party (CPC): overcome intense underdevelopment and poverty, unite the country and strengthen it to avoid imperialist humiliations, and address thousands of years of hierarchical inequalities.

The desires of hundreds of millions found unique resonance in the CPC. It was armed with an ideology of anti-imperialism and the promise of equitable development, and values venerating the common people as masters of their society.

China's revolutionary process is the story of attempts to meet these challenges differently. From 1949-76, the Mao era, China pursued a uniquely egalitarian form of socialism. It launched vast experiments in popular participation, collective social organization, women's liberation, and popular education, while maintaining steady economic growth to lay the basis for China's modern economy.

Rebirth and Reconstruction

In a report to the Joint Economic Committee of the US Congress, economist Arthur G. Ashbrook, Jr. described the China of 1949:

> Since the fall of the Manchu Dynasty in 1911, extensive areas of China had been wracked by revolution, warlordism, civil war, foreign invasion, and flood and famine. . . . Dams, irrigation systems and canals were in a state of disrepair. Railroad lines had been cut and recut by contending armies. . . . Finally, the population had suffered enormous casualties from both man-made and natural disasters and was disorganized, half starved and exhausted.[98]

Ji Chaozhu, later a prominent translator and diplomat, was from a high-ranking family in pre-1949 China. After moving to the United States during the 1930s and 1940s Ji notes that he took quickly to American food:

> The good food I'd been eating since leaving China had fueled my body's rapid growth and unleashed my natural childish energy. *I'd been starving for years.*[99]

A remarkable commentary on pre-communist China, that the child of a well-placed family was routinely starving! Joan Robinson, remarking on the predatory nature of the landlord class, noted:

> You must not think of dukes, nor yet of village squires. Here ten or twelve acres was a large estate and the landlord was not much better educated than the peasant. A large part of the income that the landlords squeezed out of the country came from usury and from cuts out of taxes that they were responsible for collecting (not to mention exactions compared to which the *droit du seigneur* [right of a

Land Reform Exhibition, 1952 (Photo: Top Photo Group)

feudal lord to have sexual relations with a vassal's bride on her wedding night—EP] seems moderate).[100]

The Chinese Revolution faced the need to stabilize and rebuild its economy on an entirely new basis. The first few years of the revolution were a period of rebuilding, but with important changes. Industries belonging to counterrevolutionaries who had abandoned their property and fled the mainland, as well as foreign companies given special privileges by capitalist China, were nationalized. So, between 1952 and 1955, private capital was eliminated, socialist planning was instituted for economic production and distribution, and the state established a monopoly of foreign trade.

Land reform, the most pressing demand for the majority of the population, was approached gradually. Large estates were broken up into smaller holdings and turned over to the peasants who worked them. Collective approaches to agriculture were set up slowly and in a skeletal form. As Chinese scholar Dongping Han notes:

Chinese peasants, under the leadership of the Chinese Communist Party and Chairman Mao, began to organize into mutual aid groups in 1950-52, lower-level agriculture cooperatives in 1953-56, and higher levels of cooperatives in 1957, consummating in the People's Communes in 1958.[101]

One reporter on the early changes in agriculture:

There were 700 families [pre-revolution—Ed.] in the village, altogether 3,000 people. The land amounted to 10,700 mow (six mow equals an acre) of which exactly half was owned by 45 landlords. Three hundred and fourteen peasant families owned no land at all. Under agrarian reform laws 551 peasants received land . . . also returned were a number of daughters and even wives taken by landlords in the settlement of debt . . . taxes were immediately reduced . . . by one-half to two-thirds.[102]

Noting further, the headman was a previously illiterate peasant, taught to read after joining the communist movement in the 1940s and elected by the villagers, not appointed as in prerevolutionary days.

As the quote above regarding the seizure of daughters and wives by landlords implies, another major effort throughout the Mao era was the focus on smashing thousands of years of feudal patriarchy and promoting the rights of women. In 1950, the first law the People's Republic passed was the Marriage Law, which legalized a woman's right to marry, or not, whomever she chose, and for all property to be held in common. One source attributes the passage of the law resulting in a rush to dissolve forced marriages that, in some areas, made up 90 percent of cases in court.[103]

That same source further notes:

In the factories . . . women got special benefits. The labor insurance laws adopted in . . . 1951, provided that women should have eight weeks' leave on full pay at childbirth . . . minimum wages for unskilled women workers were set . . . to support two persons . . . creches, nurseries and primary schools at the factories ensure that the mother is relieved of the haunting anxiety that she had in the old days.[104]

Under the slogan "Learn from the Soviet Union," the CPC approached the economy based on the model Soviet economists used. Heavy industry was given priority, with over half of all investment going to capital goods industries. Planning in consumer-goods industries was more decentralized than in the Union of Soviet Socialist Republics (USSR), and regional planning authorities had a wider purview.

Through their efforts and with significant technical assistance from the Soviets, the gross output value of all industries in China increased by 128 percent during the first Five-Year Plan, from 1953 to 1957, according to official figures. The agricultural sector only received 6.2 percent of the budget, but still managed to increase the gross output value by 24.7 percent.[105]

All this economic growth occurred alongside massive state investment in social services. The state instituted a public health program that eliminated typhoid, plague, and cholera. Educational programs were expanded, including a major campaign to end illiteracy involving fifty million peasants in 1952 and 1953.[106]

Taken as a whole, these policies were the outline of China's socialist development policy: self-reliance (breaking the chains of colonial dependency), with the goal of increasing living standards. This meant an expanded industrial sector to produce both means of production and consumer goods. An expanding industrial sector required a growing agricultural output.

China's communists endeavored to achieve both through a communist "moral economy," stressing equality and collectivity. "Raising the floor and lowering the ceiling," more efficiently harnessing the labor and machines necessary for sustained growth while allowing for the widest possible enjoyment of increased living standards.

The Context for Deepening the Revolution

Following the initial phase of the revolution, in 1957 the CPC set out to strengthen socialist approaches to development. It was a natural evolution, the strengths of collective agriculture in the Chinese context were clear enough. The benefits of economies of scale combined with egalitarian distribution already created rising living standards.

The biggest challenge for agriculture was the efficient use of water. The inefficient landlord system led to poor water usage and made rural areas susceptible to frequent droughts and floods, resulting in famines.

Cooperation among the peasants created the basis for large-scale voluntary work projects—dams, reservoirs, controlling rivers and streams—to harness the hydraulic resources and create a feedback loop into the improving economic scene.

This basic economic reality had ideological and international dimensions. Beyond reconstruction, socialist construction was a work in progress. The Soviet Union had achieved rough parity with the West, but at a high price. Chinese communists benefited from early assistance from the USSR and the broader socialist camp. This allowed a more even application of a Soviet approach.

Moving from rough parity into a qualitatively superior socialist system was paramount. However, Marx and Engels addressed the topic in a limited manner, and Lenin offered only the essential practical fragments of the first seven years of the Soviet experience. By the second half of the 1950s, the USSR and China had very different ideas about those questions.

Socialist economics is premised on conscious knowledge of the material origins of reality. Society is an answer to a most basic question: How will I (or we) survive? Capitalism offers two answers: people either work for you (capitalists) or you work for someone else (workers). The remuneration from either of the activities allows one to buy the means to hopefully survive and possibly thrive.

Socialism modifies these relationships. Rather than the anarchy of market rule, it substitutes a planned economy. The core of debates within socialist economics is: What degree of modification represents a qualitatively different social system? How does one denote the socialist road?

For Chinese communists this meant efforts to create a circumscribed role for market relationships, by formalizing the town-and-country divide, subsuming as many urban workers as possible into the state sector and as many rural peasants (and workers) into agricultural collectives. The plan prioritized producer goods and basic infrastructure that laid the basis for further economic growth: allowing agricultural productivity to increase via mechanization and scientific technique, the means to produce via machines and power, and the means to distribute more producer and consumer goods.

Productivity increases create more surplus labor. Socialist planning turns this into a plus unlike capitalism's mass unemployment. Moving workers from agriculture to industry, as well as the scientific-and-technical sector, creates more and better producer and consumer goods. The plan reveals the socialized nature of labor and the connection of one part of production to the social whole. The dependence on the collective for one's welfare allows for experimenting with moral, rather than material incentives, like voluntary labor projects.

Given that any one country's economy worked internationally, communism could be reached only after socialism encompassed the globe. This required close collaboration within the socialist camp and a united front against the imperialist camp. Disunity in the socialist camp would profoundly affect the direction of the Chinese Revolution for the next two decades.

Until 1957, the Chinese model was influenced by the Soviet experience. The Soviets, however, with no predecessors, had to find their own way in 1917. By the mid-1950s, the Soviets reached rough parity with the West, and the Soviet leadership was desperate for breathing space, which meant formalizing peace with the West.

While this was understandable, the Soviets proceeded in an unprincipled fashion. At the Twentieth Congress of the Communist Party of the Soviet Union, Soviet leader Nikita Khrushchev unleashed an attack on Stalin. Mixing truthful criticisms with serious distortions and outright lies, Khrushchev retreated from socialist economic policies using an anti-Stalinist cultural atmosphere. The attacks on Stalin masked a new focus on light industry and material inequality to address consumer and productivity issues.

In search of detente, Khrushchev used the Twentieth Congress report to introduce the new concept of peaceful coexistence, declaring it possible to have a nonrevolutionary but parliamentary road to socialism. A small group of Soviet leaders seriously contradicted, without consultation, the line of the international communist movement. They also offered an olive branch toward the main enemy of all communists—the United States—seemingly without regard to previous relationships with socialist countries.

The Soviets were willing to deny China nuclear technology to keep the peace. The atomic bomb was key to the stabilization of the Cold War. With their own bomb, China could reduce military expenditures and spend those resources on conventional needs.

The new Soviet position on the parliamentary road to socialism and related concepts of peaceful coexistence contradicted one of Marx's core tenets: "The working class cannot simply lay hold of the ready-made state machinery, and wield it for its own purposes."[107] The dizzying implications of these drastic swerves away from Marxist orthodoxy, using pseudo-Marxist phraseology, combined with differences over the international situation, escalated to such a degree that the Soviets withdrew all economic assistance from China, beginning in 1957.

In this context the CPC decided to embark on a very different direction, described by the Central Committee in November 1957:

> To carry out the technological and cultural revolution simultaneously with the socialist revolution on the political and ideological fronts; to develop industry and agriculture simultaneously with priority development of heavy industry; to develop central and local industries simultaneously under central leadership, overall planning and in coordination; and to develop large, medium and small enterprises simultaneously. To build socialism, faster, better and more economically by exerting efforts to the utmost and pressing ahead consistently.[108]

The Great Leap Forward

In January 1958 the second Five-Year Plan began. Known as the Great Leap Forward[109] it drew lessons from Soviet reversals: socialist construction could only move forward by embracing the ongoing class struggle. Breaking down distinctions between town and country, red and expert,[110] men and women, could unleash profound economic growth.

Rejecting Soviet revisionism, China's new course emphasized, as did Marx and Lenin, the importance of the masses consciously becoming the ruling class. As Lenin stated: "Communism implies Soviet power . . . enabling the mass of the oppressed to run all state affairs—without that, communism is unthinkable."[111] Through conscious planning and unleashing the energy of the masses, the foundation to transcend class society to communism could be laid. "Communism is Soviet power plus the electrification . . . since industry cannot be developed without electrification."[112]

The core of these initiatives was introducing a commune system, massive capital project mobilizations, and the expansion and shaking up of the education system.

The communization process was central to the Great Leap Forward, with the longest-lasting impact on the direction of China's revolution. For the first decade following the revolution there was some collectivization, but on a much smaller scale. In some places peasants pooled their labor on a larger scale to tackle development projects, most notably in drought-ridden Henan province. There was water on the other side of the Taihang mountains. Peasant associations brought together one hundred million peasants to carve a 1,500-kilometer canal through the mountain to water their fields, the Red Flag Canal.

Mao heard of the peasants' efforts and went to one of the villages to investigate. While participating in a meeting of a peasant association, the term commune was settled on for the new type of social organization. Once word of Mao's endorsement came out, the term swept the countryside.[113]

The communes were essentially entire rural areas that pooled together resources and labor, and functioned as the basic unit of the economy and the government. They involved thousands or tens of thousands of people. The isolation of rural life based on the single-family farm was replaced by a system where all social tasks were collectivized, from the harvest, to education, and health care. Communes across the country developed small-scale industries, supplementing large-scale capital-intensive industries still being developed. By the end of 1958, there were almost twenty-five thousand communes, embracing roughly 90 percent of the peasant population.[114]

Communes ranged in size between five and ten thousand people and swept the country so quickly that they subsumed, without completely dissolving, previous modes of peasant cooperation. Despite some chaotic management, the benefits of the commune structure quickly became apparent, especially considering:

> Of the ten biggest reservoirs in China today, the Danji-
> angkou Reservoir, Miyun Reservoir, Shisanling Reservoir,
> Xiashan Reservoir, Xinanjiang Reservoir, Lushui Reser-
> voir, Xinfengjiang Reservoir, Songtao Reservoir, Sheng-
> zhong Reservoir, and Guanyinge Reservoir, nine were

built during the Great Leap Forward. . . . During the three years of the Great Leap Forward, China made great strides in the output of steel, coal, machine tools and electricity. The increase of output over these three years accounted for 36.2 percent of China's total coal production, 29.6 percent of China's cloth production, and 25.9 percent of China's electricity generation between 1949 and 1979.[115]

The communes facilitated the quick and flexible deployment of rural labor for big agricultural and infrastructural projects. The ability of communes to conduct their own commerce with state-owned industry rewarded bursts of peasant production facilitating the mechanization of agriculture, local power production, and small-scale industrial production, freeing eighteen to twenty-one million peasants to become construction workers on major national projects and workers in new factories.

With more responsibility for education and health care, many communes rushed headlong into expansions on both fronts. Higher education and industry were hit with profound cultural shocks as workers entered universities to learn, and lecture. Administrators worked at factory benches and newspapers filled with workers' inventions and productivity-increasing improvements.

At least forty million housewives entered factories and workshops (some makeshift, set up in city streets) further severing the connection between marriage and survival. This brought expanded educational opportunities for children in new kindergartens and nurseries that sprung up in the communes and factories. Without the emancipation of women, said Mao, "socialism could not be consolidated."[116]

Despite mass mobilizations and an initial increase in production, the Great Leap Forward faced a number of organizational difficulties, many due to the still very low level of technology available for production. In the new communes, problems of organizing social life within the framework of the first attempts at economic planning caused disruptions. The benefits of collective efforts became seriously distorted. The initial rush saw a surge in volunteers of all types, a cultural phenomenon that even had high-level officials occasionally laboring on projects. Voluntary labor was organized around the principle of moral incentives. The early euphoria bred overconfidence in what this new system of collective labor could accomplish. In an effort to meet targets commandism became widespread, leading to many wasteful projects and worker-peasant alienation as the liberatory elements ebbed.[117]

Ultimately, no amount of self-reliance could make up for the more than six hundred technical aid and scientific contracts and projects lost with the Soviet withdrawal that began in 1957. Compounding these problems, massive natural disasters struck in 1960 creating both a major humanitarian crisis and an economic setback. The latter point has become one of the capitalist denunciations of communism, the centerpiece of a claim that "Mao killed millions," the capitalist world's wild claims of thirty-plus million dead between 1958-61. These estimates are, without a doubt, false.

The issue must be seen in context. First, the 1959-61 shortages and famine were the first, and last, appearance of this centuries-old scourge in socialist China. The government had managed up to that time to prevent the droughts, floods, and pests that ravaged the countryside in the Great Leap Forward years,[118] then prevented them from ever recurring, precisely because of many projects undertaken during the Great Leap. This seems like the accomplishment of centuries, rather than a black mark.

As Indian economist Utsa Patnaik puts it:

> One can much more plausibly argue precisely the opposite—that without the egalitarian distribution that the communes practised, the impact on people of the output decline, which arose for independent reasons and would have taken place anyway, would have been far worse. Further, without the 46,000 reservoirs built with collective labour on the communes up to 1980, the effects of later droughts would have been very severe.[119]

Dongping Han notes:

> In 1959, my hometown suffered a summer flood without precedent in the last hundred years. . . . In the spring of 1960, my hometown had a very bad drought. On top of that, we had another very bad summer flood. . . . But during the two years of natural disasters, we got

relief grains from the Central government, the provincial government, Qingdao City, Shanghai City and many other regions. I still remember the two dried wild vegetables shipped to us from Yunan Province. . . . For many years, my parents kept a piece of each of these wild vegetables as souvenirs of the two hardship years, and also to remember the help we got from other people in China.

People in Baoding Prefecture, Hebei Province, published a collection of memoirs titled "During the Difficult Days," which describes how, amid the severe grain shortages, people worked together helping each other, and how the local government leaders shared the hardship of the common people.[120]

Chinese mathematician Sun Jingxian, recently did an intensive study of demographic data over the relevant years. Sun unravels the complicated statistical evidence—a result of internal migration and the establishment of the household registration system—and arrives at a number of 3.66 million deaths. Not, of course, the thirty to forty-five million numbers thrown around, roughly on par with the 1928-31 famine and below the number who perished in the 1936 famine.[121]

The scale of the distortions on this question have almost wiped away any appreciation not just for the material accomplishments, but the sweep of the period in general. Han Suyin wrote, "The barrier between mental and manual labor was partly dissolved. Workers did learn philosophy; peasants did make scientific experiments. Workers lectured young students in universities; so did old peasants. Thousands of articles by workers and peasants came out in the press. Never had this been seen in China before."[122]

Nonetheless, the challenges of the Great Leap Forward required a pullback and moves to stabilize the situation. China borrowed Soviet techniques being introduced under Khrushchev, such as expanding capitalist market methods and material incentives to increase production, and increasing the authority of managers, technicians, and planners. The social goals of increasing the weight of the working class and peasantry within the relations of production became secondary.

This had several effects. Increased power for managers and technicians strengthened bureaucratic elements in the government who saw themselves as separate from the working class, though they were supposed to be in the service of that class. It also strengthened the political position of more conservative elements in the Communist Party, who used the difficulties during and following the Great Leap Forward as an opportunity to wage a political offensive against Mao and the other supporters of the socialist road.

Interregnum: 1960-1966

In the aftermath of the Great Leap Forward, China was still left with major economic challenges, at a time of growing international isolation. It was against this backdrop that the Two-Line Struggle—an expression of contending class forces after the victory of the revolution that manifested in ideological and policy debate about the most efficient method for economic development—came to the forefront of Chinese politics.

Internationally, the conflict between China and the USSR continued to escalate. The Soviets openly supported India in a border clash with China. Signing a nuclear test-ban treaty was clearly aimed at preventing China from acquiring atomic capabilities. At the same time, the West maintained an almost total isolation policy against "Red China."

Whatever the differences between them, the Chinese leadership was committed to the nationalist and anti-imperialist aspects of the revolution and chose to take a stand. They stood opposed to both powers, in their words, against imperialism on the one hand and revisionism on the other.

Chinese communists met the Soviet tensions with an ideological offensive reaffirming all the core tenets of Marxism that the Soviets seemed to be moving away from. They praised not just Marxism-Leninism in the abstract but upheld the Paris Commune and the soviets of the early Bolshevik era as their north star.[123] Collective practices in the countryside and the cities, and the experiences of the Great Leap Forward period, reaffirmed that communist theory could be practically applied.

But, how far, and to what extent was the question that underlay the real differences in the Chinese communist movement.

Ideally, socialism presupposed an advanced capitalist society where enough is already produced to fill social needs. A transitional society is as concerned with distribution as production. In China and the USSR, socialist societies were struggling to secure people's basic access to food, shelter, clothing, health care, education, and culture.

In the past, the route to the formation of modern industrial society was paved with the great crimes of genocide, slavery, and colonialism, alongside brutal subjugation of the wage laborer: the antithesis of socialist ethos. Concepts that move away from capitalism specifically and class society generally—like moral incentives—are untried and go against thousands of years of social conditioning. How much to push toward the future and how much of the past must continue to be used defined the two lines competing in Chinese communist discourse. A pullback from the Great Leap Forward years was necessary, and this empowered more conservative sectors of the party and the government.

The need for stability buoyed those seeking to bolster "the expert" over "the red," increasingly abandoning worker participation and initiative and returning to a technocracy empowering managers and skilled professionals. A worker at the Shanghai docks told two Australian researchers: "Only three periods of the week were set aside for study of any kind. . . . Prizes and bonuses were the main means of management." A textile worker added, "selfishness was encouraged." A railway worker further noted: "The Railway Ministry thought that technical problems should be solved by outside experts rather than the masses."[124]

On the other hand, the more deliberate approach to economic affairs reflected the successes as much as the failures of the Great Leap Forward. The mass agricultural projects provided more sustainable food production, and industrial and energy production expansion made an increased focus on light industry and farm products realistic. It all added up to a quick economic recovery between 1962-65.

While the CPC's left wing was on the defensive, areas of innovation continued in the countryside and factories. In 1960, workers at the massive Manchurian steel complex at Anshan, after two days of interacting with the Chairman, promulgated a statement of principles—the Anshan Constitution—for economic management. Emphasizing "that workers' enthusiasm in production should be based on workers' consciousness . . . and that the production would contribute to the long-term interests of the working class as a whole."[125]

These principles underlying factory management during the Mao era became the bedrock of Mao's socialist road approach. Economist Hao Qi observed during his fieldwork: "Workers I interviewed recalled the factory life in the Maoist era, they *always* mentioned that workers were the masters of the factory."[126] Mao's main opponents within the party were Liu Shaoqi, a veteran member with the top post of chairman of the People's Republic of China (PRC), and Deng Xiaoping, a fellow veteran of the Long March. They were the main proponents of reversing the course that Mao had initiated in the Great Leap Forward that still lived in places like the Anshan steel works. Mao called Liu, Deng and their cothinkers capitalist roaders.

The Cultural Revolution

These issues were the seeds of conflict leading to the Great Proletarian Cultural Revolution. The capitalist roaders argued for a defensive policy, saying that in light of international isolation and underdevelopment, some scope of market activities—highly circumscribed—was essential to socialist construction. Whatever the moral implications, the capitalist roaders believed the "old methods" were rooted in real material realities and could not be bypassed.

The socialist road made an offensive argument: ideology was a material factor in production. Central planning lifts the fog over the socialized nature of production. With the consciousness of one's role in the process of social reproduction, the masses can be motivated through the moral incentive of collective achievements. Taking the best parts of the Great Leap Forward as their starting point, they saw a clear connection between revolutionary consciousness and productivity and growth. In the final analysis, socialist growth could only come by relying on the masses themselves, the mass line.

Dongping Han describes the social upheaval of the Great Proletarian Cultural Revolution,

> The Great Proletarian Cultural Revolution tried to build a real democracy. It empowered the ordinary Chinese people to write big character posters to criticize their leaders . . . most Chinese officials had lifestyles similar to those of ordinary people. They lived in houses similar to those of ordinary people. Their children went to the same schools as

other Chinese people. They went to work on bicycles like everybody else. Production team leaders were elected by peasants and worked with peasants in the field every day.[127]

By 1965 the pot was on boil, and in 1966 the struggle broke out into the open. Despite the heavy economic differences between the two "roads" or "lines," the opening salvo of the Cultural Revolution came in the form of literary criticism. In November 1965, the Shanghai newspaper *Wenhui Bao* published an article criticizing a play called *Hai Rui Dismissed from Office*.[128] The article accused the playwright of using the story of a Ming Dynasty official to criticize the party and Mao. Since the play's author was the deputy mayor of Beijing, it was clear the criticism was not for academic debate. The criticism's targets were capitalist roaders—found in the party's upper echelons. The critique's author, Yao Wenyuan, was based in Shanghai, a center of socialist road sympathizers, and where Mao based himself in the years leading up to the Cultural Revolution. The criticism said of the party leadership in Beijing, "no needle can penetrate it; no water can percolate through it."[129]

The choice of literary criticism, and culture more generally, as the opening salvo of the struggle, was not arbitrary. Culture and knowledge are the foundation of socialist values and morals underpinning the transitional process. If these fields remained mired in worship of China's feudal past, its customs, and hierarchies, how could they strengthen the foundations of a society premised on entirely different values? Further, the relatively cloistered world of the cultural apparatus was a reflection of how education, especially higher education, was a province of a fairly narrow elite rooted in the industrial-party-political bureaucracy. Their own children had better preparation, and thus privileged access, to educational and managerial opportunities. How could culture, or science for that matter, serve the goals of a new socialist society if it didn't open up these opportunities to those from the revolutionary classes?

The first rumbles of the great movement about to sweep the country for four years appeared in the world of education, in the spring and summer of 1966. Starting at Beijing University, students put up big character posters, publicly posting political pronouncements denouncing school and party leaders, and demanding more participation for working class and peasant students.

In one rural county a high school student and a physics teacher put up their first big character poster in June, criticizing authorities for focusing too much on standardized test scores and excluding working-class students from educational opportunities. On June 6 an entire middle-school class wrote to the Central Committee demanding the abolition of college entrance exams, suggesting instead that middle-school graduates be embedded in the factories and communes, and chosen by the masses to matriculate to higher levels.[130] The youth revolt was underpinned, initially, by the People's Liberation Army (PLA). Under Lin Biao,[131] who was an ally of Mao's and the minister of national defense, the PLA had become a center of socialist road views and practice—abolishing ranks and becoming a key part of large mass projects. Tasked with politicizing young people, the PLA network and publications inspired initial groups of organized students that emerged in 1966, under the name Red Guards.

Capitalist roader elements used the party apparatus to stifle the student revolt, even setting up some official Red Guard units. They tried to direct student anger toward lower levels of officialdom. However, when schools closed in June, larger numbers of students began to parade and demonstrate, as well as put up big character posters. They noticed the differences between what party work teams were saying and what PLA and Mao-aligned publications were putting out. They saw the need to challenge even the highest level of officials.

In Beijing, Mao had the same message to the Politburo: "You must lead the fire of the Great Cultural Revolution toward yourselves. . . . You yourselves must fan the flames to make them burn you."[132] In this context, Mao made the decisive move to appeal to the masses themselves and the revolutionaries in the party, to expose the revisionism that risked China's socialist path, wherever it was, even at the highest levels.

First, Mao replied to students at Tsinghua University Middle School, published in a Red Guard publication: "You say it is right to rebel against reactionaries; I enthusiastically support you. I also give enthusiastic support to the big-character poster of the Red Flag Combat Group of Beijing University Middle School which said that it is right to rebel against the reactionaries."[133] Then on August 5, *Beijing*

Review carried Mao's first big character poster to all of China and the world: "Bombard the Headquarters," calling on people to attack anyone promoting revisionism, no matter their position.

The Great Proletarian Cultural Revolution was officially announced in August 1966. The Central Committee of the CPC adopted the "Decision Concerning the Great Proletarian Cultural Revolution," also known as the "Sixteen Points." It read in part:

> Although the bourgeoisie has been overthrown, it is still trying to use the old ideas, culture, customs and habits of the exploiting classes to corrupt the masses, capture their minds and endeavor to stage a comeback. The proletariat must do just the opposite: it must meet head-on every challenge of the bourgeoisie in the ideological field. . . . At present, our objective is to struggle against and overthrow those persons in authority who are taking the capitalist road, to criticize and repudiate the reactionary bourgeois academic "authorities" and the ideology of the bourgeoisie . . . and to transform education, literature and art and all other parts of the superstructure not in correspondence to the socialist economic base, so as to facilitate the consolidation and development of the socialist system.[134]

Students, then teachers, put up wall posters by the thousands, identifying particular school and government officials as capitalist roaders. Millions of students across the country took part, forming many organizations with the name "Rebel" in them, denoting their fealty to the struggle against reactionaries and to Chairman Mao's ideas. Groups of students began to move around the country linking up with worker and peasant rebels.

The task was aided by publication of a small bound volume of quotes summarizing the essence of Marxist-Leninist tenets and Mao's observations. Compact and short, the bite-sized style was aimed at mass readership in a country where illiteracy still existed, making oral transmission of information key. Easy to remember and repeat this "little red book" was an instrument of political education and a framework for political action anyone could use and understand.

Not content to just put up posters, students began to openly denounce and replace school and government leaders. Although Mao and his allies in the Central Committee were the minority inside the leadership, their political ideas corresponded to the political thinking of millions of students and teachers, and growing layers of the working class.

US and other Western media glorify protests or dissent within a country with a socialist government. Championing the cause of workers and students against communist bureaucrats is the norm. The smallest protest in Cuba, the former Soviet Union, or in China today, is promoted with great fanfare in the corporate media.

Not so for the Cultural Revolution. When millions of young people—later workers and peasants as well—participated in the Cultural Revolution against party leaders, who preserved privilege and inequality, the big-business media treated them as dangerous lunatics. The ruling class understood the meaning of this movement. It was aimed at overcoming the problems of Chinese society by taking the next step toward communism—rather than by returning to capitalism.

What was unique about Mao and Lin Biao's campaign was that it appealed "over the heads" of party leaders directly to the masses. This contrasts with Leninist norms of democratic centralism practiced by communist parties around the world. According to these norms, Mao should restrict his criticisms to inner-party discussions during congresses and plenums. To the masses, the party would maintain a united position. This was a criticism of the revisionist Soviet leadership in calling the Cultural Revolution ultraleft.

These criticisms ignored the fact that Lenin and the Bolsheviks adopted democratic centralism as the best way for a revolutionary political program to be put into practice. Formalistically justifying a reactionary political program or practice has nothing in common with revolutionary Marxism.

The Rebellion Builds

Throughout late 1966 and into 1967, more people were drawn into the Cultural Revolution. Society was cleaved in two and rebel associations sprang up everywhere. In rural Jimo County there were rebel associations among and within artists, the Bureau of Material

Supplies, Jimo Tractor Station, and the Jimo Hardware Factory. Rebel associations in villages and factories linked up with similarly minded people in other villages and factories, often banding together in large fronts, aiding each other in mass struggles. Old leaders were denounced and replaced by new ones from the working class and the youth. Mass meetings in government buildings and in the factories became routine. Officials were made to produce documents proving their guilt of various crimes and were subject to public questioning and ridicule. Many crimes were revealed and mass justice often resulted in the humiliation, beating, and occasional death of those guilty of serious transgressions.

The proliferation of mass associations that crossed occupational, geographic, and party membership lines dramatically altered the balance of power in society. The self-organization of workers and peasants was given official sanction. Through alliances with like-minded supporters up and down the party-government-economic axis, they had not just sanction, but the ability to make changes, unleashing another wave of mass initiative.

In January and February of 1967, workers in Shanghai launched what may have been the most advanced initiative of the entire Chinese Revolution. Revolutionary workers' organizations seized the city's two main newspapers and declared the Shanghai Commune.[135] It was explicitly modeled on the Paris Commune of March to May 1871, the first time the working class seized and held political power. The Shanghai Commune was based on dozens of rebel organizations of tens of millions of workers, many of which elected their leaders—with provision for automatic recall. Commune officials took their posts after a democratic procedure in their own rebel fronts.[136] For a time it controlled city affairs, marginalizing official government structures while drastically shrinking the size of the bureaucracy, by transferring some army and police functions to an armed militia, and abolishing sectors of the state security forces. Declaring that only those guilty of "murder, arson, poisoning" and "sabotage of state secrets" would be arrested, as well as using "people's mediation committees," they resolved most civil and criminal disputes at the grassroots, outside the formal courts.[137]

The events in Shanghai were the high point of the "January Storm," and similar experiments sprang up around China, with Shanghai-like features and their own innovations, one of which, the Revolutionary Committees, gained favor among Maoists in the party center. The big character posters were connected to mass associations, and created a peoples' media with investigative power. This combined the ability to air grievances with the ability to question officials through mass pressure. In South River Village in Shandong province, big character posters prompted revelations of illegal grain hoarding and a possible murder by a deputy party secretary, years before.[138]

The process was aided by students traveling to and from Beijing to meet with revolutionary forces, determined to pierce the veil that party and state officials in isolated rural areas hid behind, and confirm the correctness of the revolt against officials. Dongping Han notes about his own Jimo village:

> In 1966 one group of twenty rural youth between four-teen and sixteen years old left their rural middle school in Jimo County on foot for Beijing. They were the first young-sters from their villages who had ever ventured beyond the county town. . . . Holding a red flag, they were determined to walk all the way to Beijing. . . . They . . . stopped for the night at various reception centers. . . . At these reception centers they met students from other places and discussed the developments of the Cultural Revolution with them. As they saw the world, and exchanged ideas with others, they felt politically empowered.

> However, these direct interventions of the working class and peasantry were complicated and nuanced. Many sympathetic to those denounced as "capitalist roaders" as well as existing state structures, formed their own mass associations. Society was rent top-to-bottom with various "storms."[139]

Mass meetings with denunciations and self-criticisms became ubiquitous. Some capitalist roader factions directed mass campaigns at lower-level officials to protect those higher up. New anarchist factions emerged, labeling many administrative tasks as positions of power to be attacked. They interfered with China's aid to the Vietnamese liberation struggle. Red Guards and the military clashed in places, resulting in some massacres and huge numbers of arrests of rebel Red Guards.

Under these circumstances, Mao's leadership group retreated from the commune-style state. Fearing that the growing struggles were leading to a large-scale violent confrontation, they reasserted control of the bureaucracy to a certain extent. The result was a compromise of sorts. Mao now promoted three-in-one combination committees, sometimes called the triple alliance. Political power was exercised through committees with representatives of the mass youth, workers' and peasants' organizations, members of the People's Liberation Army, and Communist Party cadres. These Revolutionary Committees had emerged in several rebel-controlled areas and offered a blend of approaches.

They reaffirmed the importance of a centralized state structure, carving out key roles for the party and the PLA as stabilizing influences. At the same time they promoted a formal governing role for the mass associations growing out of the mass democratic upsurge. This was in some ways a retreat from the workers' and peasants' democracy that emerged from the Cultural Revolution's mass mobilizations. It marked a step away from the commune-style of state beginning to emerge. On the other hand, it recognized the importance of a strong state structure in a hostile imperialist world and maintained critical social compacts of the revolution, including the country's territorial integrity.

Revolutionary Committees represented a shift in the balance of power, in favor of workers and peasants, in daily life. It gave hundreds of millions the opportunity to observe, indict, supervise, temporarily detain, punish, replace, and reprimand those in managerial and leadership positions. It formalized the rights of workers to determine key elements of their own working conditions and remuneration structures.

Stabilization and Revolution

The energy released by the Cultural Revolution impacted US activists who spent time in China during the period:

Modern China represents an enormous effort to create a truly egalitarian society. We saw evidence of this everywhere; in the practical emphasis on learning from the peasants and the workers, the relatively narrow range of pay differentials, the stress on simple living and the similarity of living standards, the allocation of housing, the commonly expressed belief that to serve the people is the greatest goal to which an individual can aspire.[140]

These changed circumstances led to stabilization without abandoning the core of the Cultural Revolution. Economic growth and easing China's international isolation created a strong foundation to change day-to-day power relations for workers and peasants, and institutionalized new practices.

For instance, within the state-mandated eight-grade wage policy, workers set the wages for each grade and decided together which workers were in which grades. In the Wuhan Sewing Machine factory workers voted on grades in open discussions in their shop units for who rates what grade, at intervals determined by the workers themselves. The Revolutionary Committee approved the results or sent them back for more discussion. Workers judged income decisions based on a range of factors they developed, which were most universal in their consideration of seniority.[141]

In Zhengzhou's No. 3 Textile Mill workers emphasized "political consciousness and skill." At Laoyang Tractor Works they added a definition for "high political consciousness": over-fulfilling tasks, good relations with coworkers, and a clear understanding of "political struggle as class struggle." Kairan Coal Mine workers valued the type of work done, offering more to those working underground. Wuhan Iron and Steel Works gave bonuses to those who worked in high places.[142]

Communes had a similar system. Income distribution was based on work points that entitled one to proceeds from collective work. The Cultural Revolution moved away from a pure work-points system to collective decision-making around income distribution. A glimpse from a story from Red Star commune outside Beijing:

A woman in one village rose up to challenge the leadership which had changed her daily work points from 10 to 8 points just because she had gotten married. Under these pressures, the question of women's equality was put to the whole commune for discussion. Women asked, "when we do the same work as men, why do we get less work points?" The men replied, "because you are weak and we are strong!" The women retorted, "OK, for everyday in the year that our work depends on strength, we'll agree to getting less, but for the other days we must get the same! Let's see who plants rice the fastest!" So they organized a rice planting competition. The women outstripped the men by far. The men conceded. From then on, women got 10 points, the same as men.[143]

This led to further transformations:

One evening we heard criticisms of party secretary Wong from such and such a village. In the discussions he had agreed that since women also worked in the fields, men should help with the housework. He announced proudly, "I'll do anything needed at home. Except for emptying the pot and changing the baby's diapers, I'll do anything." "What's wrong with men emptying the pot?" came a sharp woman's voice over the loudspeaker. Men everywhere started. Emptying the pot! One day there was suddenly a great commotion beating of drums and clashing of cymbals. What's going on? A group of young couples were getting married. The bridegrooms had all volunteered to become members of their wives' families instead of the wives becoming the members of the bridegrooms' families. Within a few months, with all these changes going on, villages started training young women as electricians, carpenters, tractor drivers and even mule drivers! "Women hold up half the sky!"[144]

Revolutionary Committees overseeing the process were chosen in various ways. Some spots were filled by the PLA and the party, centralized structures that were the key link between the center and the grassroots. It included the popular participation that began in the rebel mass associations.

In the Wuhan Sewing Machine Factory workers made nominations from their shop units to be voted on by the whole factory. In Jimo County the initial revolutionary committee was elected from associations of poor and middle peasants along with delegated committees from rebel mass organizations. Workers at a locomotive manufacturing plant in Tangshan created something called the "two-times up, two-times down" process of selecting. Workers nominated candidates from their areas and it was sent to the existing Revolutionary Committee. The committee then held a discussion and gave their response, and the process was repeated. After that, all involved decided on the representatives by consensus without a vote. In a Shanghai neighborhood, neighborhood committees were made up of elected representatives from street committees as well as party and PLA representatives.[145]

As mentioned above, access to education was a cornerstone of the Cultural Revolution a key element of the massive transformations. In Jimo County, for instance, schools sprung up across the county. Almost every village established primary schools, where there had been almost none, enrolling 98 percent of elementary-age children by 1973 and 99 percent by 1976. Villages banded together using their own materials and labor to build 130 middle schools. In 1962 Jimo County had eighty-eight high school students. By 1976 it had just over thirteen thousand.[146] Teachers were drawn from two main pools: local people with education or relevant experience, and teachers of the old government system sent from cities to teach in rural areas, often areas they had migrated from.[147] In Evergreen Commune, in the countryside around Guangzhou, the situation was similar:

> The comrades in the People's Militia . . . are sent here to teach about military questions. . . . The commune workers give lessons on agricultural mechanics, the brigade accountant teaches mathematics. Health officials teach hygiene.[148]

Visiting a teachers' college in the mid-1970s, famed Black writer John O. Killens, related the following concerning their curriculum:

> A young Chinese woman, nineteenish, came up to me and asked: "Are you an Afro-American?" I said, "Yes." She told me, "I am getting my degree in English." She was a very pretty woman, self-assured without arrogance, with olive-smooth and silky skin and deep, dark, slanting eyes. . . . She said to me, "In our class, we are studying a novel that has been translated from English into Chinese about

Black soldiers in World War II. . . . While she was telling the story, I was having a weird feeling that I had either read this story before or had heard of the incident. I asked, "What was the title of the book?" She said, "The title of the book is *And Then the Thunder Was Heard*." I said, "You don't mean *And Then We Heard the Thunder?*" "She said, "Yes . . . of course! That's it. Have you also read the novel?" "Yes," I admitted, "I have also read the novel. I also wrote the novel. . . ." "You're John Keerins! You're Keerins," she shouted softly, her face glowing with excitement. I was equally excited. . . . The incident gave me an eerie feeling. For here I was halfway around the world, and there are upwards of 30 million Afro-Americans in this world and 800 million Chinese people. And you could get all kinds of odds in decadent Las Vegas against this coincidence occurring. But after I got to Beijing I met with Chinese publishers and found that they had also published my novel *Youngblood*, as well as *Black Man's Burden*. They had published many of the books of Dr. W. E. B. DuBois. They had also published Langston Hughes, Paul Robeson, Robert Williams and other Afro-American writers.[149]

The Detroit judge and esteemed Black jurist George Crockett went to China specifically to look at the judicial system noting, among other things, that:

To my considerable amazement, I found that *crime just isn't a problem* in China today. . . . What I actually encountered was a scarcity of formal legal proceedings, with the people themselves tending to solve both civil and criminal conflicts. . . . Everything is in the hands of the people themselves. They make the decisions on a level they can understand completely, without the legal trappings that Americans have come to associate with the dispensing of justice."[150]

Health care was rapidly expanded across rural areas. In 1965, just before the Cultural Revolution, Mao had taken aim at the Ministry of

Health, criticizing them for focusing almost exclusively on the cities, a practice ended by the late-1960s. A Dr. Ling, from a Shanghai hospital notes: "In 1968, ten thousand worker-doctors were sent from Shanghai into the rural zones." By the early 1970s they had trained twenty-nine thousand "barefoot doctors."

The barefoot doctors became world famous as news about the Cultural Revolution seeped into the world. Peasant youth from the communes were trained to treat basic illnesses and in traditional plant-based medicine. Combined with traveling doctor brigades and an expanding professional health-care system, 350 million peasants with almost no access to health care in 1965 had at least rudimentary care by the early 1970s.[151]

The later period of the Cultural Revolution also sought to embed, via decree and habit, a spirit of narrowing the gap between mental and manual labor. Emphasis was put on pre-Cultural Revolution students and professionals to work in the countryside. Millions were sent to rural areas with important, but mixed results. This became a controversial part of the Cultural Revolution. The basic thrust was to take relatively privileged people and place them in decidedly less privileged circumstances, many doing manual labor for the first time. The process was carried out somewhat haphazardly, so experiences varied from fairly ideal, clearly defined periods, to ones where people were essentially forgotten. For professionals, the "sent-down" processes were linked to a reeducation movement, and became a weapon of factional warfare.

On the other hand, there are numerous examples of a profound opening of the world on both sides. Current Chinese President Xi Jingping says many of his ideas and characteristics were formed in his time living in a cave in rural Liangjiahe. In Jimo County, sent-down teachers ended up being a windfall for rural villages establishing middle schools. Dr. Wan, a Shanghai surgeon, tells of gaining rich experience practicing medicine and working in the fields from 1967-69.[152]

Some of the displacement was the result of a type of affirmative action adopted when colleges and universities started to reopen in 1970. In Tsinghua University, for example, 45 percent of students were selected from the factories, 40 percent from the communes, and 15 percent from the PLA. This was a big change prior to 1968 when 60 percent of the students came from nonworking-class origins.[153]

Commune and village leaders were also integrated into rural work schedules alongside their official duties. Dongping Han remembers:

> Farmers still remember Ding Qichao, the head of the Jimo County Revolutionary Committee, riding his bike on the dirt road in the fields . . . stopping once in a while to work and talk with farmers.[154]

Workers were directly integrated into the planning process. State-planning targets were developed in consultation with party committees, and then transmitted to the shop floor for worker feedback and collective agreement between the various levels as to the final targets. Emphasis was placed on providing time for workers in factories to study during the work day, the "philosophy in the factories." The phrase "grasp revolution, promote production!" became famous in this regard.

The concept was that active participation in production required not just formal debate and decision-making, but the broader principles of political economy and philosophy at play, demystifying their own role in society and its relationship to the construction of socialism and maintaining national independence. By combining abstract theory within the rhythm of the work week, alongside the cross-fertilization of technical experience with managers working on the shop floor, productivity would be spurred. Understanding the ways their collective efforts led to the improvement of society as a whole worked as a moral incentive.

For example, in the pre-Cultural Revolution days, the port of Tianjin lacked strong enough cranes to handle many commercial vessels from abroad. The workers were studying Mao's *On Contradiction*, specifically the dialectical unity of opposites, whether conditions internal or external to a relationship are decisive in the outcome. Fortified by Mao's explication of the revolutionary concept that the internal conditions are always decisive: "In a suitable temperature an egg changes into a chicken, but no temperature can change a stone into a chicken,"[155] the workers embarked on a project to build a new, heavier crane. Working with scrap for eight months, five of those living at the job, the workers successfully delivered the new crane, donating it in honor of the Ninth Congress of the Chinese Communist Party. This philosophical crane was topped with a red flag adorned with a portrait of a smiling Chairman Mao, and emblazoned with a slogan: "Be Self-Reliant."[156]

In the countryside mass projects incorporated popular initiatives, and new rural leaders solidified their standing by working alongside peasants and rural workers. As a set of interviews with peasants from a rural village relates: "The reason why leaders worked harder during the Cultural Revolution was simple. Common villagers would not tolerate lazy leaders. If leaders did not work, villagers refused to work as well. . . . If leaders did not work hard, villagers would elect someone else . . . in the year-end election."[157]

In 1968 South River Village started a large irrigation project, collectively voted on by the villagers and with their input into engineering and technical aspects. "At night, villagers who worked on other projects . . . all came out to put in a couple hours of work. . . . School teachers, students, and local government employees all came to help. . . . Government worker, Chu Jiying . . . said that she, like others, volunteered to work at the projects . . . because it was an honorable thing to do."[158]

Political campaigns stressing high theory/politics played a role in asserting agency by rural Chinese peasants. Han explains:

> The campaign to criticize Lin Biao and Confucius . . . was a good example of this. What had Lin Biao and Confucius to do with common villagers? Why should common farmers be concerned with power struggles among the top leaders or with what an ancient philosopher said more than two thousand years ago? For thousands of years, Chinese villagers lived very apolitical lives, seldom getting involved in politics. The fact that Mao and other Cultural Revolution leaders saw the need to involve common villagers, most of whom were illiterate . . . was in itself revolutionary and democratic. . . . A central point of the campaign . . . was that rural people were not stupid. . . . It promoted . . . that the masses are the motive force of history. . . . In this sense the campaign served to help common Chinese villagers discover their dignity.[159]

Every revolution includes a degree of chaos, arbitrary action, and excess. This was certainly the case during the Cultural Revolution. But much of the reporting was distorted both by the world bourgeoisie who feared the revolutionary mobilization, and later by the Chinese leadership who were the targets of the mass mobilizations. But, economic statistics don't reflect a disaster or even stagnation. With agricultural production increasing above levels preceding the crisis that began in 1959, Chinese industry continued moving forward. They developed a method for producing durable, low-alloy steel; they produced thirty times as much in 1969 than 1965. By the early 1970s, China achieved self-sufficiency in cotton, wool and silk, and was the world's largest producer of cotton. Tractor production doubled in the first seven months of 1970, as compared to that in all of 1966. China was producing more than a thousand different models of agricultural machinery—three hundred more than in 1966.[160] Author William Hinton describes the advances:

> In 1959 peasants had to come up with 116,500 kg. of wheat to buy a 75-horsepower track tractor. By 1979, this fell to 53,500 kg. In 1950 it took 1.6 kg. of wheat to buy 1 kg. of fertilizer. By 1979 0.5 kg. of wheat would buy 1 kg. of fertilizer. In 1960, it took 35 kg. of wheat to buy 1 kg. of pesticide. By 1979, this had decreased to 5 kg.[161]

Mobo Gao summarizes the overall advances in the standard of living since the victory of the revolution:

> Measured by social indicators such as life expectancy, infant mortality and educational attainment, China, especially urban China, in the Mao era had already forged way ahead of most market economies at similar income levels and surpassed a number of countries with per capita incomes many times greater. By the late 1970s, China stood up as a nuclear power, able to defy the bullying of capitalist superpowers, a country that had satellite technology and became the sixth largest industrial power in the world."[162]

Each commune had its own schools, hospitals and medical centers. "Barefoot doctors" were trained in basic medical procedures, spreading access to basic health care to millions. In 1973, 50 percent of all medical students were women. By 1971, 90 percent of women had jobs working outside of the home.[163] As a Quaker delegation to China noted:

Throughout our visit we saw concrete evidence that China's women were sharing equally the task of building the new China. While most of the women work on the land, we also saw women wearing hard hats on construction jobs, operating overhead cranes in factories, at work on lathes. We saw school girls repairing electrical equipment, women pulling heavy loads of produce or of baggage or of driving trucks. All these soon became familiar sights. . . . We found Chinese women proud of their new status. They like to quote Chairman Mao that men and women are equal and that anything a man can do a woman can also do.[164]

Breaking Out of Isolation

Underlying China's relative stability was its move out of international isolation, and ultimately into a de facto alliance with Western imperialism on several fronts. The first opening however, came from the left. Then, as now, tens of millions of people were hoping for alternatives to the soulless consumer capitalism of the West and breaking the bonds of former colonial dependencies.

As news of the Great Leap Forward, and its most hopeful remnants, started to seep out from behind the imperialist blockade of "Red China" since the late 1940s, many started to see the country as a beacon of hope for communism and anti-imperialism. These early reports—combined with the CPC's bold stands internationally, reaffirming the most revolutionary core of Marxism, and translating that into a development strategy—created excitement and wonder.

Harry Belafonte, in an interview with Helen Rosen (a notable Paul Robeson confidante) said in this regard:

When Alassane Diop, Guinea's former Minister of Communications, came back from a visit to the new China in the early 50s, he told me that the city of Shanghai was clean and beautiful, that its citizens had a decency and spirit unequaled anywhere else in the world. I asked myself how a nation devastated by war and riddled with hunger, disease, and illiteracy was able to order the lives of eight hundred million citizens. I erupted into an insatiable curiosity about China."[165]

In 1967 Guinea officially declared itself socialist, launched its own cultural revolution of sorts, and adopted a range of organizational and developmental policies drawing influence from the Maoist group in China. Major splits took place across communist movements around the world as Chinese critiques of imperialism and issues within the USSR, along with their socialist policies, gained more adherents.

In France and Italy powerful communist movements experienced splits that saw Maoism gain millions of adherents and penetrate into politics and culture. In the United States Chairman Mao was an icon and teacher for the Black Panthers, and a generation of tens of thousands of young people radicalized by the Vietnam war and determined to renew the US communist movement. China counted among its prominent friends artist Harry Belafonte, French filmmaker Jean-Luc Godard, prominent Black American writers Alice Childress, Amiri Baraka, and John Killens, plus Joan Robinson, the doyen of social democratic economics in the United Kingdom.

China won friends not only through its domestic policies but by taking stands in support of anti-colonial and anti-capitalist resistance everywhere. China worked with Tanzania and Zambia to build the eleven-hunded-mile TAZARA railway. It went from the Port of Dar Es Saalam to the copper belt of Zambia. China helped break Zambia's dependence on white-ruled South Africa and Rhodesia and opened up new trade possibilities for Tanzania.

Chinese revolutionary strategy spread to Mozambique and Zimbabwe where FRELIMO and the Zimbabwe African National Union respectively received decisive training and aid, and adopted Mao's people's war strategy for their liberation struggles. China hailed the Naxalbari peasant uprising in India in 1967 that turned into a powerful new communist current. China associated closely with the US Black liberation movement. Shirley Graham DuBois was a prominent friend of China and a link to longstanding relationships between China and Black radicals like Paul Robeson, Vicki Garvin, and W.E.B. Du Bois.

China hosted Black militant Robert F. Williams for several years and developed ties with the Republic of New Afrika, of which he was provisional president. Imari Obadele, central leader of the Republic of New Afrika, described Lin Biao, head of the PLA, as one of their leading supporters, a sign of how deep China's desire to promote revolution around the world penetrated.[166] China's stance was buffeted by the traditional belief in the communist movement that socialism was a world system, and not a one-nation or multi-nation affair. Progress may be uneven, so all the more important to promote revolution and grow the forces of the socialist camp. Tolerating isolation and siege as a condition for existence was not a virtue. Ultimately, no major revolution emerged at a scale to tilt the world balance of power. China faced a challenging situation. While its developmental achievements were by any measure spectacular, they still represented progress from a low base. Ironically, the successes of socialist construction revealed challenges and possible limits.

The Twilight of the Mao Era

The struggles of the Cultural Revolution took place against a backdrop of practicality. Unlike the Great Leap Forward, the approach to growth was realistic in scale. Growth was steady but not far ahead of population growth. A key part of development is productivity, being able to do more with less effort. This allows societies to support larger numbers of mental workers; people teaching and learning, in the science and technology sectors, cultural workers, and such.

Under capitalism the struggle for profit promotes productivity. Capitalists only increase their share of the overall value produced in society at the expense of workers. It can be done in two ways. First, make people work harder and longer for less. Second, apply science and technique and do more with less. There are clearer limits to the former than the latter. Capital seeks to displace people with machines, with contradictory consequences: it lowers the cost of living and allows a society with more people doing essential but nonproductive tasks. But it also results in mass unemployment that grows structurally over time.

Socialism posits the nonexploitative method of consciousness building and moral incentives. On a project-by-project basis, this can have fantastic results as we have seen, systematizing gains on a country-wide level. William Hinton relates that in conversations with key economists, they told him that taking the communes as a whole:

Some 30 percent of the village collectives were doing well, another 30 percent were doing very poorly, and in the middle the remaining 40 percent could be said to be neither thriving nor collapsing, but that nevertheless they confronted many problems.[167]

Clearly, as the 1970s wore on stagnation set in, a natural result of an economy that can't grow much more than its population growth. There was enough development in rural areas to create labor surpluses, but not adequate ways to expand the possibilities. This led to steady low growth and shadow unemployment as workers nominally attached to workplaces or communes worked inefficiently and sometimes infrequently, while still drawing on the social wage of housing, health care, education, and so on.

The most obvious solution to this problem was for China to break its isolation from the USSR, the US, or both. Easing relations would bring greater access to advanced technologies and capabilities as well as the possibilities of using China's growing labor surplus to collective advantage. At different times China took one-sided approaches to both questions, ultimately reversing previous stands.

China opened relations with the United States in 1972, and became a permanent UN Security Council member. While this did not need to be mutually exclusive of improved relations with the Soviets, China carried it out that way. Doing what they had accused the USSR of a decade before, China secured its future at the expense of the Soviet Union. China improved relations with the West but accommodated itself to elements of imperialist foreign policy by supporting the apartheid-backed UNITA rebels in Angola, the fascist Pinochet regime, Filipino dictator Ferdinand Marcos, and the Afghan Mujahadeen.

Domestically it seemed Mao wanted a mixed approach to addressing the challenges of building socialism. For instance, in 1973, Chen Yonggui was elevated to the Politburo. A formerly illiterate peasant leader, he'd become a national hero leading one of the most successful commune efforts in China. He spent time working on the commune as a top official and attended the highest diplomatic functions in peasant dress, a powerful example of China's egalitarian revolution.

The year 1973 was when Mao brought Deng Xiaoping back into the national leadership.[168] Clearly Mao, sending signals to his closest supporters "not to form a Gang of Four," wanted to promote a balanced approach. After all, as William Hinton points out, the 30 percent of successful communes contained 240 million people, easily the greatest global effort at cooperative economics ever mounted. The efforts of the successful areas had something to contribute to improve the performance of other less successful areas. However, these attempts had to give way to retreats. Deng and those around him had innovative ideas about how to use market forces within a socialist system that deserved a hearing. Nonetheless, Mao's supporters did form a Gang of Four who waged an aggressive war against those they labeled capitalist roaders.

As the economy continued to stagnate, the initiative shifted to Deng and market socialist forces several years after Mao's death in 1976. This ultimately led to a total reversal of post-1949 socialist policies.

Balance Sheet

Modern China is inconceivable without the conquests of the Maoist period from the revolution until 1976. As Chinese scholar Mobo Gao reminds us:

We have to remind the world that it was during the Mao era that the average Chinese life expectancy rose from 38 years in 1949 to 68 years by the 1970s. During the same period, the literacy rate increased dramatically and rural health improved dramatically, so much so that it prepared for millions and millions of skilled and healthy workers in the post-Mao period economic expansion. . . . Socialist China's GNP grew at an average annual rate of 6.2 per cent between 1952 and 1978. . . . Measured by social indicators such as life expectancy, infant mortality and educational attainment, China, especially urban China, in the Mao era had already forged way ahead of most market economies at similar income levels.[169]

The Mao period attacked not privileges, but privilege as such, involving hundreds of millions in the crucial national questions and the quality of their daily lives. The economic results speak for themselves. The Great Leap Forward and the Great Proletarian Cultural

Revolution successfully applied socialist principles to national development. China went from a society where the majority of people were guaranteed none of the means of survival to a society where everyone was guaranteed the basics of life—food, clothing, shelter, health care, and education—*a reversal of thousands of years of inequality in less than a generation.*

These chapters of class struggle in China remain a reserve source for the workers of China and the world in the struggle for revolutionary socialism.

GLOSSARY
OF NAMES

The 1898 Reformers: Kang Youwei (1858-1927), Liang Qichao (1873-1929), and Tan Sitong (1865-1898)

These three men were the main leaders of the Hundred Days Reform movement in the summer of 1898. The Guangxu Emperor tried to carry out reforms to revive the fortunes of the Qing dynasty, and appointed Kang, Liang, and Tan to be special counselors. The reforms were aborted by the reactionary Empress Dowager Cixi in September.

Kang Youwei was a traditional Confucian scholar who developed new interpretations of Confucian teachings. He wrote a book called *Confucius as Reformer* that validated innovation and modernization. In 1895, in the wake of China's defeat in the First Sino-Japanese War, he led a protest movement among candidates for the imperial examinations to petition the imperial government for reforms. After the failure of the 1898 Reforms, he spent much of his time outside China. He continued to advocate reform of the imperial state, and called for a constitutional monarchy. He opposed the creation of the Republic in 1912 and remained a monarchist the rest of his life.

Liang Qichao was a close associate of Kang. A traditional scholar, he wrote an important book on the intellectual history of the Qing dynasty. He was also active in the 1895 protests. After 1898, he too spent most of his time outside China, much of it in Japan or in Shanghai, where he was safe from imperial power in the foreign concessions. He became a publisher of radical newspapers and journals, though, like Kang, he remained committed to the idea of a constitutional monarchy.

Tan Sitong was the highest-ranking official of the 1898 Reforms. His father was a provincial governor, and he passed the imperial examinations with a high result. Tan wrote essays presenting what he saw as modern elements within Chinese cultural traditions. His most important book was *Renxue* (The Study of Benevolence), in which he explored both scientific and political ideas. In 1898 he was made a grand secretary, the highest position in the civil administration. When the Empress Dowager's anti-reform coup took place, while Kang Youwei and Liang Qichao fled to Japan, Tan chose to remain in Beijing, where he was arrested and executed with seven others in September. He believed political change required the sacrifice of martyrs, and embraced that fate to promote his political vision of a new China.

All China Federation of Trade Unions (ACFTU)

Originally established in 1925, during the First United Front between the Communist Party of China (CPC) and the Guomindang, the ACFTU became the official national organization of the labor movement when the People's Republic was founded in 1949. It is an umbrella organization representing all workers in China. It was suspended during the Cultural Revolution when its functions were subsumed within the Revolutionary Committees, but was reinstated in 1978. Today the ACFTU includes more than 1.7 million trade union organizations, with a total membership of over three hundred million workers. The ACFTU advocates for workers' interests within the CPC and the government of the People's Republic. In a socialist state there is no sharp delineation between the government and the economy, so the ACFTU works to represent labor with both public authorities and private enterprises. It is headquartered in Beijing, where it publishes the *Worker's Daily* and many other journals and magazines, as well as other media.

All-China Women's Federation (ACWF)

The ACWF was founded on April 3, 1949. It is a mass organization that unites Chinese women of all ethnic groups and all walks of life, and strives for their liberation and development. The mission of ACWF is to represent and uphold women's rights and interests and to promote equality between women and men. It maintains a very active publishing program, producing magazines, books, electronic media, and other materials in both Chinese and languages of minority communities in China. It also prints in foreign languages, including the English language monthly magazine *Women of China*. The ACWF is very active in promoting women's rights in the workplace and the role of women in education and the professions. The slogan "women hold up half the sky" has long encapsulated the commitment of the Chinese Revolution to recognizing the equality of women and men.

The Boxers

The term Boxers is a shorthand reference to a mass anti-imperialist movement that arose in Shandong province in 1898 and spread across northern China over the following two years. It culminated with the siege of the Legation Quarter in Beijing in the summer of 1900, where the foreign diplomatic community had been established. The movement was grounded in a martial arts discipline and was properly known as the Society of Righteousness and Harmony *(Yihetuan)*, and the name Boxers was used because of their martial arts practices. The Boxers opposed the activities of German missionaries in Shandong and the presence of the German concession in Qingdao, on the south coast of the province. They called for the expulsion of foreigners from China and the ending of the system of treaty ports and domination by Western imperialism. The Boxer rebellion was suppressed by a joint military expedition of eight imperialist powers, which invaded China and occupied Beijing in late July 1900. Many rebels were killed and the Qing government was burdened with a huge indemnity to pay the costs of the invasion, and to punish the country for resisting its humiliation.

Central Cultural Revolution Group (CCRG)

The CCRG was an ad hoc body formed in May 1966 as the Cultural Revolution was getting underway. Its membership fluctuated over time but was always focused on Mao Zedong and his closest advisors such as Chen Boda, Yao Wenyuan, and Zhang Chunqiao. The CCRG played the leading role in articulating a critique of bureaucratism within the CPC and the threat of a return to the capitalist road. Members of the CCRG were involved in the most dramatic moments of the Cultural Revolution, such as the Shanghai Commune and the Wuhan Incident. The CCRG ceased to function as a political center after the Ninth Party Congress in April 1969, while several of its members fell from positions of influence. Others, especially Yao, Zhang, and Jiang Qing, carried on the effort to oppose bureaucratic tendencies through the early to mid-1970s, when they came to be referred to as the Gang of Four, along with Wang Hongwen.

Chiang Kai-shek (1887-1976)

Chiang Kai-shek was born in Zhejiang province. He came from a family of landowners and merchants, but decided to pursue a military career. He met Sun Yatsen while enrolled at a military academy in Japan. He joined Sun's Revolutionary Alliance and in 1911 was a regimental commander in the fight against the Qing forces. After the Bolshevik Revolution in 1917, he studied in the Soviet Union briefly, but became strongly anti-communist. After Sun Yatsen's death, Chiang Kai-shek became the dominant figure in the Guomindang (GMD), and succeeded in uniting more and more of China under GMD control while becoming increasingly hostile to the communists. In April 1927 he split with the communists, carried out a massacre of cadres and labor activists in Shanghai, and destroyed the united front.

Through the early 1930s Chiang launched a series of attacks on the CPC base area in Jiangxi, aimed at exterminating the Red Army and the communist leadership. This led to the Long March and the shift of CPC headquarters to Yan'an. In 1936 Chiang was kidnapped in Xi'an by the warlord Zhang Xueliang, who was frustrated that while Chiang was focused on the communist threat, Japan seized Manchuria. His kidnapper forced Chiang to make peace with the communists and establish a Second United Front against Japan. After the war with Japan ended in 1945, the Guomindang lost the Civil War to the communists in 1949 and Chiang fled with his Republic of China government to Taiwan, where he ruled under martial law until his death in 1976.

Chinese People's Political Consultative Conference (CPPCC)

The CPPCC developed as part of the struggle for New Democracy during the Chinese revolutionary civil war from 1945-49. The Communist Party argued that China needed a new form of democratic participation, that it needed and wanted the input of other patriotic groups and individuals in the process of winning the revolution and creating a new political system. In September 1949, on the eve of the proclamation of the People's Republic, the CPPCC convened and passed the Common Program, which provided the guidance and organization of the new government for the first years of the People's republic of China (PRC). Until 1955 the CPPCC functioned as the legislative branch of the government, after which that role was assumed by the National People's Congress (NPC). The CPPCC continues to function today, bringing together organizations and individuals in a process of deliberation, and providing recommendations for policies and legislation to the NPC. Both groups meet on a coordinated five-year cycle. Between full sessions of the organization a National Committee carries on the work of the CPPCC, consulting with and advising the leadership of the CPC, and the government of the PRC.

Deng Xiaoping (1904-1997)

Deng Xiaoping was born in Sichuan province into a family of the Hakka community, a distinct identity group within the overall Han ethnicity, but distinguished by being the descendants of Chinese who had migrated south during the Tang dynasty, after initial settlement of the region several centuries earlier. He attended school in Chongqing. After graduating in 1919 he joined a group of young Chinese who went to France to take part in a work-study program. In France he worked for a while at the Renault factory outside Paris, where he met Zhou Enlai and grew acquainted with the French Communist Party. Deng was one of the earliest members in the French branch of the Communist Party of China. In the mid-1920s he spent time in Moscow. He returned to China and became a political secretary in the Red Army by 1929.

Deng allied himself with Mao Zedong's views of the revolutionary agency of the peasantry, the agricultural proletarians who they saw as the natural allies of the urban workers. He rose in the leadership when Mao became the key leader at the Zunyi Conference early in the Long March. Deng played active roles in the Long March and during the Anti-Japanese War and the Civil War. After Liberation he was the party secretary in charge of Southwest China. He was increasingly involved in economic policy and in the development of science and technology.

During the Cultural Revolution Deng was associated with Liu Shaoqi and labeled a "person in power in the party taking the capitalist road." He was removed from office and spent time working in southern China. In 1972 he returned to a leading role with responsibility for the "Four Modernizations" in science and technology, and represented China at the United Nations several times. He was again removed from power briefly, then once again brought back late in 1974, only to be demoted yet again early in 1976.

After the death of Mao Zedong in September 1976, Deng was deeply involved in the internal debates about the future of socialist development. By November 1978 he emerged as the effective leader of the CPC and the government, though he never accepted the position of chairman or president. He became the main architect of the new policies of reform and opening to the outside world. He remained the key figure in the reform efforts and reinvigorated the process of reform in 1992, after the disruptions associated with the events of 1989 in Beijing. He sought to use the mechanisms of the market to develop China's productive economy, in order to reach a state of material abundance which would enable the implementation of socialist distribution, while recognizing that this involved serious risks and contradictions and could only succeed with the guidance and leadership of the Communist Party. Deng Xiaoping died in 1997.

Empress Dowager Cixi (1835-1908)

Daughter of a minor Manchu official, she was betrothed to her cousin Ronglu before an invitation to meet the young Xianfeng Emperor (r. 1851-61) changed the course of her future. He named her "the Yi Concubine" and she gave him a healthy male heir in 1856. Xianfeng's reign was marked by the Taiping Rebellion and the Second Opium War (1856-60). He trusted the Yi Concubine as an advisor.

In 1861 Xianfeng died and Cixi became co-Empress Dowager, along with another of Xianfeng's consorts. She carried out a coup against the group of regents who gained sway over the emperor in the months before his death. Her six-year-old son, who had become the Tongzhi Emperor (r. 1862-74), declared that all decisions would go through the Empresses Dowager. From then on Cixi was the real power behind her son.

During the 1860s efforts at reforming the dynasty, known as the Tongzhi Restoration, took place. After the Tongzhi Emperor died, Cixi arranged for her nephew to become the Guangxu Emperor (r. 1875-1908). In 1898, when Guangxu attempted to implement the Hundred Days' Reforms, Cixi placed him under house arrest and executed several of his advisors. In 1900 she put her support behind the Boxer Rebellion but was forced to make conciliatory gestures toward the foreigners after its defeat. In 1908 both Guangxu and Cixi died within a few hours of each other.

The 'Gang of Four': Jiang Qing (1914-1999), Zhang Chunqiao (1917-2005), Yao Wenyuan (1931-2005), and Wang Hongwen (1935-1992)

The term "Gang of Four" was a negative label applied by their enemies in the CPC to this group of political figures associated with the later stages of the Cultural Revolution.

Jiang Qing was a film actor in Shanghai before moving to Yan'an in the late 1930s to join the revolution. She married Mao Zedong, and for many years maintained a low profile, not participating in political affairs. When the Cultural Revolution began, though, her background in theater led her to become active in the development of revolutionary operas and other new cultural productions. Her closeness to Chairman Mao gave her a special prominence and influence.

Zhang Chunqiao was a writer and literary critic in Shanghai in the 1930s. He joined the CPC and became an important official in the Shanghai propaganda bureau after liberation. He developed a critique of the bourgeois right, the idea that new bourgeois elements could emerge within China if bureaucratic tendencies in the party and government were not checked. This brought him to the attention of Mao. When the Cultural Revolution began Zhang emerged as a strong leader of the movement in Shanghai. He was elected to the Shanghai Commune in early February 1967, and was part of the delegation that met with Chairman Mao, leading to the formation of the Revolutionary Committee that replaced the Commune at the end of February.

Yao Wenyuan was also a literary figure in Shanghai. He became a personal aid to Chairman Mao, and in 1965 wrote the first essay criticizing Wu Han's play *Hai Rui Dismissed from Office* as an attack on Chairman Mao and the Great Leap Forward. This essay became the opening salvo in the Cultural Revolution. Yao was also elected to the Commune, and was one of the delegates who met with Mao in Beijing.

Wang Hongwen was a factory worker in Shanghai when the Cultural Revolution began, and quickly became a prominent activist in the mass movement. In February 1967 he was elected to the Shanghai Commune, and later joined the leadership of the Revolutionary Committee. He rose quickly in the Central Cultural Revolution Group, and became a vice-chairman of the CPC.

After the Ninth Party Congress in 1969, the mass movement phase of the Cultural Revolution ended. Jiang, Zhang, Yao, and Wang carried on the promotion of a radical-left critique of cultural policies and the overall problem of bureaucratization in the party. Zhang and Yao wrote important articles setting out their views, while Jiang continued to work on the creation of new cultural forms and activities. They strongly opposed the rehabilitation of Deng Xiaoping in 1972.

After Chairman Mao died in September 1976 the political center within the party shifted against the radical positions advocated by these leaders, and in October all four were arrested. They were tried and sentenced to prison. Jiang Qing was released from prison for health reasons, but died soon after, in 1999. Wang died of cancer in 1992. Zhang was released from prison for health reasons in 1998, and died in Shanghai in 2005. Yao was released in 1995 and lived quietly in Shanghai, writing works on Chinese history, until his death from diabetes in 2005.

Guomindang (GMD)

The GMD was the party of the nationalist bourgeoisie. It developed out of a coalition of revolutionary organizations, the *Tongmenghui*, dedicated to the overthrow of the Qing dynasty and the imperial system in the first decade of the twentieth century. Sun Yatsen was the leader of the GMD from its founding until his death in 1925. When the Qing was deposed in 1911-12 the GMD was positioned to lead a new republican government, but that project was derailed by the warlord Yuan Shikai. The GMD was marginalized as a regional power centered in Guangdong province on the south coast. After the Bolshevik Revolution in Russia the Communist International sent agents to China to work with local political forces opposed to imperialism and to promote the development of a revolutionary movement. They worked with Sun and the GMD and helped reorganize the party along democratic centralist lines. The GMD entered into a United Front with the newly formed Communist Party from 1923-27. But after Sun's death in 1925 the GMD's new leader was the reactionary Chiang Kai-shek, who launched a bloody split with the CPC in April 1927. The GMD became increasingly corrupt and embroiled with organized criminal groups. From 1927-49 Chiang led the GMD in an effort to destroy the CPC, but this failed and in 1949 the GMD remnants fled the mainland to the island of Taiwan. They had prepared for this retreat with the repression of local political forces and the imposition of martial law in 1948. The GMD held power on the island until 1987, and since then has alternated control of the local government there with other parties. Today the GMD has become a force for reconciliation with the central government in Beijing, opposed to the calls for independence on the part of some elements on the island.

Hong Xiuquan (1814-1864)

Hong Xiuquan was from a Hakka family, a Chinese ethnic group in southern China, living near Guangzhou (Canton). He studied for the imperial examinations but failed repeatedly, as did most candidates. Eventually the stress of these repeated frustrations led to a nervous breakdown, during which Hong saw visions of two men who talked with him about mysterious things. Once while in Guangzhou for the exams, Hong was given Christian pamphlets by a Western missionary, which he put in a drawer in his study. After recovering from his breakdown, he reread the pamphlets and came to believe the men he'd seen in his vision were God and Jesus, and that he was the younger brother of Jesus, meant to bring the new religion to China.

Hong began to preach a version of Christian ideology emphasizing social equality and justice. Many people in southern China suffered from economic dislocations caused by the opening of China to Western imperialism after the First Opium War (1839-42). Hong's message gained many followers, and he became the leader of a movement known as the Heavenly Kingdom of Great Peace (*Taiping Tianguo*). By the 1850s this became a great rebellion against the Qing dynasty, to establish a new system and overthrow the old society. The Taipings marched north and established a capital at Nanjing. The Qing suppressed and finally defeated them in 1864. Hong was killed when Nanjing fell to imperial forces.

Hu Jintao (1942-)

Hu Jintao, like Jiang Zemin, was a technically trained specialist, in his case a hydraulic engineer. He was from Jiangsu province, where he grew up in a poor family, though he traced his ancestry to a famous Ming dynasty general. He graduated from Qinghua University in 1964 and joined the Communist Party the same year. He worked as an engineer on water projects in Gansu province through the late 1960s and served as party secretary in various units.

When Deng Xiaoping launched the policies of reform and opening, Hu was drawn into more party work. He became active in the Communist Youth League, an important incubator for future leaders. He served as party secretary for Guizhou province in southern China and then in the Tibet Autonomous Region through the 1980s and early 1990s. In 1992 he was put in charge of the party secretariat in Beijing and was named to the Political Bureau, the CPC's top policy-making group.

In 2002 Hu was elected to succeed Jiang Zemin as CPC general secretary and as PRC president. He took up these positions early in 2003 and served two five-years terms. His administration oversaw the continuing growth of the economy, with 10 percent expansion of GDP in many years. He engaged with Guomindang political leaders in Taiwan province in a period of reconciliation, when new travel and communication links were opened, and when investment on the mainland from Taiwan expanded. Hu maintained the international posture of "biding time and building capacity" as adopted by Jiang Zemin. In 2012, Xi Jinping was elected to succeed Hu in the leading offices of the party and the government.

Hu Yaobang (1915-1989)

Hu was a Hakka, a member of a distinctive ethnic group in southern China. His family were poor peasants in Hunan, and Hu's early life was quite harsh. He had no formal education, but taught himself to read. He left home to join the Communist Party when he was fourteen, becoming a full member when he was eighteen. Hu endured many hardships as one of the youngest participants in the Long March. He spent time as a prisoner of the GMD forces, but escaped and reached the Yan'an base area with more than a thousand other Red Army soldiers. He became close to both Deng Xiaoping and Peng Dehuai during the Yan'an years (1937-45) and the Civil War (1945-49).

After liberation Hu became a leader of the Communist Youth League (CYL), a key factor in his political career. Many future leaders in the party began their careers in the CYL, so this was an important nexus of relationships. During the Cultural Revolution Hu was removed from office, then restored to party work several times, in sync with the fortunes of Deng Xiaoping.

When Deng became the key leader after the death of Mao Zedong, he raised Hu Yaobang to the top of the party leadership. Hu was made general secretary of the CPC in 1980, a position functionally replacing the role of chairman, which was formally abolished the next year. Through the 1980s Hu was one of the strongest advocates for all-out marketization, along with Zhao Ziyang. As economic and social contradictions emerged by the mid-1980s many senior party figures saw Hu as having gone too far in pushing for marketization, and he was removed as general secretary in 1987, though he retained his seat on the Political Bureau, the top party policy-making group. He was replaced by Zhao Ziyang.

Hu died of a heart attack during a meeting of the Political Bureau in April 1989. In the wake of his death there were student demonstrations in Beijing by those who sought to expand their position in political affairs in line with Hu's support for all-out reform. These demonstrations developed into the occupation of central Beijing and a movement that became increasingly anti-government, calling for the overthrow of the CPC and the PRC. When negotiations broke down this resulted in the intervention of the People's Liberation Army (PLA) to restore order in the capital on June 4.

The International (Third/Communist International)

The International, known alternatively as the Third or the Communist International, sometimes abbreviated to the Comintern, was established in Moscow on March 4, 1919, at a meeting of delegates from the main revolutionary parties of Europe and the United States. The Bolshevik Revolution had taken place with the expectation that revolutions would soon follow in the major European industrialized countries. Communist uprisings did briefly seize power in Berlin, Budapest, and in a few other urban centers, but these were rapidly crushed by reactionary forces. The formation of the International was intended to provide an organization to support and encourage revolutionary activities in countries around the world. In the first years of its activities a strong focus on building movements in the colonial and semicolonial world was developed, with China becoming a primary area of concern and engagement. The International sent advisors to China, notably Maring and Otto Braun, who helped organize the Communist Party there and also worked with the Whampoa Military Academy. The International struggled to develop a viable strategy for revolution in China, maintaining an orientation based on the weakness of the industrial proletariat and the need to collaborate with the patriotic national bourgeoisie, even after Chiang Kai-shek's anti-communist coup in April 1927. As the CPC adopted its peasant strategy in the 1930s, the views of Mao and the Chinese leadership diverged from those of the International, and the Soviet Union continued assistance to the GMD government until 1948. The International was formally disbanded during World War II, as the Soviet Union participated in the alliance with the United States, France, and Britain against fascist Germany and Italy.

Jiang Zemin (1926-2022)

Jiang is representative of the third generation of leadership within the CPC and the government of China. Born in Jiangsu province, he was the adopted son of Jiang Shangqing, who was killed fighting the Japanese during their occupation of northern China in the early 1940s, and is considered a national hero. Jiang Zemin graduated as an electrical engineer in 1947. He joined the Communist Party during his college years. After liberation he worked as an engineer in the Stalin Autoworks, then in the Changchun First Automobile Works. He became involved in party activities, and over time his main work shifted from technical engineering to political and administrative tasks. His career path is seen as typical of the emergence of a "technocratic" leadership within the CPC.

Jiang became mayor and party secretary of Shanghai in 1985. When political unrest developed in the spring of 1989 Jiang managed the situation in Shanghai through meetings with student protesters, and prevented the kind of disruption that developed in Beijing. After the end of the occupation of central Beijing in early June, and the dismissal of Zhao Ziyang as general secretary of the CPC, Jiang was appointed to take over that position. He held the highest offices in the party and the government from 1989 to 2003, though in practice Deng Xiaoping continued to be the strongest influence on policy until his death in 1997.

Jiang's period of leadership saw the resumption of the policies of reform and opening and the period of rapid growth in the economy that began in 1992. He developed the theory of the "Three Represents"—part of an effort to broaden the political base of the CPC. The "Three Represents" was the idea that the party should represent the development of China's productive economy, the quality of its advanced culture, and the interests of the vast majority of the people. Its most practical expression was opening party membership to some capitalists, if they were considered patriotic and contributed to the development of the productive economy. This was a contentious decision and disputed by various elements within the CPC.

Jiang followed Deng Xiaoping's advice that during this period of economic development through engagement with the global capitalist system, China should keep a low international profile. Deng argued that China should "bide its time, develop its capacities," meaning China should not emphasize its socialist agenda in its relationships with foreign countries, especially the United States. Jiang oversaw the negotiations that led to China joining the World Trade Organization in 2001.

After Deng's death in 1997 Jiang presided over the process of leadership transition, and in 2003 he stepped down as general secretary of the CPC and PRC president. He was succeeded by Hu Jintao.

Lin Biao (1907-1971)

Lin came from a merchant family in Hubei province. As a student he became active in radical politics, and became involved with a branch of the Communist Youth League in high school. He enrolled at the Whampoa Military Academy in Guangzhou, run by the GMD and the CPC during the First United Front, when Chiang Kai-shek was the director and Zhou Enlai was an instructor. During the turmoil of the late 1920s Lin was involved in various actions and wound up at the Jiangxi Soviet, where he rose to be an important military leader. He played a leading role in the Long March, and was a divisional commander in the first year of the war with Japan. In 1938 he was shot in a confused encounter with other Chinese forces, and contracted tuberculosis while recovering from his wound. He was sent to Moscow where he served as a liaison with the Soviets.

When Japan surrendered in August 1945 Lin was transferred to the northeast, where he became commander of Red Army forces working with Soviet units in the area. He played a leading role in the Civil War (1945-49), and led the fighting in 1948-49 that drove the GMD armies out of the mainland. He was seen as one of the best field commanders of the PLA, along with Peng Dehuai.

Lin maintained a low profile during the 1950s, due to health issues. In 1959, as a result of the political conflicts at the Lushan Plenum, he was named minister of defense, replacing Peng Dehuai. In that role he oversaw the revitalization of the PLA's political department, and supervised the compilation of the *Quotations from Chairman Mao* as a text for political education within the army. This became the famous "little red book" during the Cultural Revolution.

Lin was a very outspoken supporter of Mao Zedong during the Cultural Revolution, and was named as his "close comrade in arms and successor" in the party constitution adopted at the Ninth Party Congress in April 1969. But relations between Lin and Mao deteriorated as Mao revised his views on the primary contradiction in global affairs and moved toward an opening to the United States. The exact nature of events in 1970 and 1971 remain unclear, but in September 1971 Lin was killed in a plane crash in Mongolia.

Liu Shaoqi (1898-1969)

Liu Shaoqi was from Hunan province, like Mao Zedong. He came from a moderately well-off peasant family. He received a modern education in Hunan, and then was sent to Shanghai. He was drawn to socialist activities there, and in 1921 went to Moscow to study at the University for the Toilers of the East, the Soviet school for Asian students. He joined the CPC and returned to China in 1922. He became deeply involved with labor activism and emerged as one of the leading figures in labor organizing for the party during the United Front. In 1927 he was elected to the central committee and made head of the party's labor department.

In 1932, Liu moved to the Jiangxi Soviet base area, and in 1934 took part in the early phase of the Long March. After the Zunyi Conference in February 1935, he was sent back to Shanghai to oversee the party's labor work in the "white areas," the area of China controlled by the GMD. In 1937 he again shifted his location, this time to the new CPC headquarters at Yan'an. He rose in the party leadership, and after liberation was a vice-chairman of the CPC and served as chairman of the Standing Committee of the National People's Congress in the 1950s.

After initially supporting the Great Leap Forward, Liu became increasingly critical of Mao's leadership. At the Lushan Plenum in August 1959 Liu was chosen to replace Mao as president of the PRC. When the Cultural Revolution got underway in 1966, Liu became the focus of criticism of bureaucratism in the party, and was labeled the chief person in power taking the capitalist road. He was removed from his leadership positions in 1968, and died of cancer the next year.

Mao Zedong (1893-1976)

Mao Zedong was born to a family of farmers in Hunan province. By the time of the May Fourth Movement in 1919, Mao was writing on such topics as nationalism and gender inequality in China. In 1921, after participating in early Marxist reading groups, he became a founding member of the CPC. During the First United Front Mao became an organizer for the Guomindang Peasant Bureau in Hunan, where he developed his theories on the revolutionary role of the farmers in China. In 1929 he established the Jiangxi Soviet as the main base area for the CPC and the Red Army. Mao participated in the Long March and became the effective leader of the CPC at the Zunyi Conference in early 1935. He was formally named chairman of the party in 1945.

During the Yan'an years (1936-45) Mao wrote important works on Marxist theory and strategy, creating a system of revolutionary thought adapted to the conditions of Chinese society. These included his main philosophical works, *On Contradiction* and *On Practice*, as well as his talks on cultural issues at the Yan'an Forum on Art and Literature. He also developed his ideas on military affairs and political strategy, constantly evolving these as the revolution proceeded.

On October 1, 1949, Mao announced the founding of the PRC. As chairman of the CPC, Mao was a key player in the struggle between the two lines in the party and clashed with leaders of the "pragmatist" tendency within the party, such as Liu Shaoqi and Deng Xiaoping. He was the main promoter of the Great Leap Forward in 1958-60, which faltered due to factors including bad weather, bureaucratic distortion of production figures, and the split with the Soviets in 1959. At the Lushan Plenum in August 1959 Mao had to step back from day-to-day leadership, and relinquished the position of the PRC chairman to Liu Shaoqi.

In 1966, Mao launched The Great Proletarian Cultural Revolution, aimed at rooting out the bureaucratic tendencies of party cadres and renewing the links between the party and the masses. Mao was influenced by the deterioration of his country's relationship with the Soviet Union. The Sino-Soviet split was followed by the start of a more friendly relationship with the West, and Mao famously met with American president Richard Nixon in 1972. Mao's death in September 1976 was followed by major changes within the CPC and the government of the PRC, leading to the shift to the policies of reform and opening under Deng Xiaoping's leadership after 1978.

May Fourth Movement

After the end of World War I the victorious allied powers convened a peace conference at Versailles, outside Paris, where they set about establishing the world order they wished to maintain into the future. While the rhetoric of self-determination for all peoples had been deployed during the war, especially by the American president Woodrow Wilson, in reality the allies never intended this to apply to their own colonial subjects. China had supported the allies during the war, sending tens of thousands of workers to France. Japan also joined the allied side, and occupied German territorial possessions, including Qingdao and its hinterlands in China, and several islands in the western Pacific Ocean. China expected that with the war's end its territory would be restored, but instead the allies granted Japan continued possession of Qingdao. News of this arrived by telegram on the morning of May 4, 1919, and triggered student-led demonstrations in central Beijing that day. These were attacked and repressed by military forces of the warlord government, with many students arrested. A nationwide movement opposed to Japanese imperialism quickly developed, with strikes, boycotts, and demonstrations in cities around the country. The May Fourth Movement exposed the hypocrisy and bankruptcy of Western liberal democracy and the real intentions of the imperialist powers to continue their domination and exploitation of people in the colonial world. This became a major contributing factor in the quest for a new path to national salvation and the turn to Marxism and socialism as inspiration for the emerging revolutionary movement.

Peng Dehuai (1898-1974)

Peng was born into a poor peasant family in Hunan, and received a minimal formal education. In his early teens he worked as a manual laborer, and when he was sixteen he became a professional soldier. He served in local forces in Hunan, then with the Guomindang army during the First United Front, and finally joined with the communists after the split of April 1927 and the failure of the Nanchang and Autumn

Harvest Uprisings later that year. He rose in the ranks of the Red Army, first in the Jiangxi Soviet, then on the Long March, and on through the War of Resistance to Japanese Aggression (1937-45). He was one of the key commanders in the Civil War (1945-49), leading major operations in the northeast. After liberation, when the war in Korea broke out in 1950, Peng argued in favor of Chinese intervention, and became field commander of the Chinese People's Volunteers from November 1950 through the middle of 1952, when he returned to China and took over the day-to-day management of the Central Military Commission.

In 1954 Peng was named minister of defense. He worked to modernize the PLA, and was close to Soviet military advisers in China during the period of close cooperation between China and the USSR in the 1950s. He was an advocate of professionalization of the military, seen by Mao and his associates as moving away from the model of the PLA as a people's army.

When the problems of the Great Leap Forward emerged, Peng was a leading critic of Mao Zedong. At the Lushan Plenum in August 1959 Peng circulated a letter he drafted among other party leaders, without showing it to Mao himself. This led to a confrontation between them, resulting in a political compromise. Mao withdrew from the day-to-day management of party affairs and stepped down as president of the People's Republic, while Peng resigned as minister of defense and was replaced by Lin Biao. Peng lived in semi-retirement in a garden compound on the grounds of Beijing University until 1965, when he was appointed to direct a new initiative for developing defense industries in China's southwestern region.

Peng was criticized during the Cultural Revolution and was the subject of several struggle campaigns because of his views on the nature of the military, having advocated a professionalized army rather than the model of the people's army upheld by Mao and Lin Biao. His health deteriorated and he died in 1974.

Red Guards

The Red Guards were the young rebel activists, primarily students but with some participation by young workers as well, who burst upon the political stage in August 1966. This was in response to Mao Zedong's call to "bombard the headquarters" and the pronouncement that "it is right to rebel against reactionaries" in his big character poster at Beijing University. Through the fall of 1966 a series of mass rallies of Red Guards were held in Tiananmen Square at the center of Beijing, each of which included about a million young people. The Red Guards were not a unified organization with a coherent political line, but were rather a broad spectrum of groupings that often differed radically in their ideas and actions. Many Red Guards embarked on revolutionary travel ventures, visiting sites associated with the revolution, an activity encouraged by the leaders of the Cultural Revolution and supported by granting free passage on the country's railroads. Factionalism among the Red Guards led to violent conflicts in numerous instances, and the movement became chaotic, leading to the decision in 1968 to encourage young urban students to go to the countryside to work alongside peasants in their villages in order to deepen their understanding of the need for developing a new socialist society. Many of these "sent down" youths remained in rural areas until the late 1970s, when most returned to their original homes in cities as the political orientation of the CPC shifted after Mao's death. Some former Red Guards chose to remain in the villages where they had developed close ties with the local community and found new ways to contribute to the creation of the new China.

Self-Strengthening Movement

In the nineteenth century, after its defeat and humiliation at the hands of British imperialism, the creation of treaty ports, and the broader opening of the country to Western exploitation, some Chinese political leaders sought ways to resist and counter the power of the industrialized capitalist core. Beginning in the 1860s reform efforts were undertaken, which included establishing a government office to manage relations with Western states and the founding of an arsenal and shipyard to develop modern military capabilities. By the 1870s and 1880s this became the Self-Strengthening Movement, as the Qing imperial state tried to raise its ability to stand up to foreign encroachment. The movement faced dual obstacles. On one hand the imperialists sought to weaken China and to slow its process of modernization by only selling obsolete weaponry or ships to the Qing. Within the imperial government there were many officials and members of the Manchu nobility who opposed these efforts, believing they would undermine their own power and privilege. In the end the

Self-Strengthening Movement had some achievements, but it was unable to give China sufficient military or industrial power to hold off imperialist domination. China was defeated in wars with France in 1885, and with Japan in 1894-95, which had embarked on its own course of rapid modernization in the 1860s.

Sun Yatsen (1866-1925)

Sun Yatsen was born in Guangdong, near Hong Kong. He came from a peasant family, and was educated at a missionary school in the Kingdom of Hawai'i where his older brother was living, before getting his medical degree in Hong Kong. Sun developed a revolutionary, anti-Qing fervor and founded the Revive China Society in 1894. His goal was to make China into a republic, and he opposed figures such as Kang Youwei and his followers, who advocated a constitutional monarchy. After staging the first of several unsuccessful military uprisings, Sun was forced to flee China. While in London, he was kidnapped by imperial officials and held at the Chinese Legation, making him internationally famous. In 1905 Sun became head of the Revolutionary Alliance. He used the *Three Principles of the People* to articulate the ideology of the Revolutionary Alliance. The principles are often translated as socialism, nationalism, and democracy.

After the 1911 revolution ended the Qing dynasty, Sun Yatsen was made provisional president of a new republican China. He gave up the position to the Qing general Yuan Shikai and focused his efforts on transforming the Revolutionary Alliance into the Guomindang political party, also known as the Nationalists. When Yuan turned on the GMD and pushed them out of government, Sun Yatsen had to leave China again. From then on his mission was building up his party and preparing to unite China, which had descended into warlord rule, under GMD control. Under his leadership the GMD entered a united front with the communists from 1923-27, and received significant support from the Soviet Union and the Communist International. In 1925 he died of liver cancer.

Three-in-One Combination (Revolutionary Committees)

The Three-in-One Revolutionary Committees emerged in early 1967 and quickly became the form of political organization and administration adopted by cities, provinces, and other units across the country. The classic example, which became the model for other places, was established in Shanghai at the end of February. The Shanghai Commune had been set up on February 5 by mass organizations of workers after more than a month of agitation in the city, overthrowing the authority of the Municipal Party Committee and the civil government of Shanghai. Drawing on the experience of the Paris Commune of 1871, the idea was to create a system of direct proletarian democracy. After discussions with Mao Zedong in Beijing, however, it was agreed that it was still necessary for the Communist Party to play a leading role in the struggle for socialism, and that the People's Liberation Army needed to be relied on to provide security and stability. The Three-in-One combination of representatives of the party, the army, and the masses, was embodied in a new Revolutionary Committee that took over the administration of China's largest and most industrialized city. This model of unifying the masses under the leadership of the CPC and the PLA was soon seen as the best path forward, and by the beginning of 1969 Revolutionary Committees had been established nationwide.

Xi Jinping (1953-)

Xi Jinping is the first top leader of China born after liberation. His father, Xi Zhongxun, was a vice-premier and vice-chairman of the National People's Congress. Xi Zhongxun was criticized and sent to work in a tractor factory at the start of the Cultural Revolution. Xi Jinping was sent to the countryside in 1968, as were most Red Guards from the cities. He spent seven years in Shaanxi province. He joined the Communist Youth League after several applications. In 1974 he became a regular member of the Communist Party. He went to Qinghua University as a worker-peasant-soldier student in the final years of the Cultural Revolution, and graduated in 1979.

In the 1980s and 1990s Xi served in a series of provincial party posts. From 1998 to 2002 he studied Marxist theory at Qinghua University, and earned a doctorate in law degree. During the decade of Hu Jintao's leadership, Xi ran the Central Party School in Beijing, directed planning for the 2008 Olympics, and oversaw policy for the Hong Kong and Macau Special Administrative Regions.

Xi Jinping was elected to succeed Hu as general secretary of the CPC and as PRC president in 2012, and assumed those offices in

2013. He was reelected in 2018. Under his leadership China moved away from the low-profile posture of the previous twenty years, and has become more self-confident and assertive in international affairs, even as the United States has become more hostile and aggressive toward China.

Xi has reemphasized the party's place in guiding economic development and the pursuit of a socialist future for China. He has promoted the study of Marxism, and has called for "completing the original mission" of the revolution. He launched the Belt and Road Initiative, a program of aid and investment in the development of global trade infrastructure bringing countries in Asia, Africa, Latin America, and even Europe into a new era of relations with China and one another.

Zhao Ziyang (1919-2005)

Zhao came from a wealthy landlord family in Henan province. He joined the Communist Youth League when he was thirteen, and became a member of the CPC when he was nineteen. He held a number of administrative positions in local and provincial level party units through the 1940s, and served for a while in the Red Army. After liberation he became a regional party leader in southern China. He was involved in the Great Leap Forward in Guangdong province, and became an advocate of the reinstatement of market forces to promote production during the crisis years of 1960-61. In 1965 he became party secretary for the province.

Zhao was criticized during the Cultural Revolution. He was dismissed from his party office and went to work in a machine factory for four years. In 1971 he was rehabilitated and appointed to an administrative position in Inner Mongolia. He rose in party offices through the 1970s. In 1975 he became party secretary for Sichuan province, where he experimented with economic reforms allowing a greater role for the market in agriculture.

When Deng Xiaoping began the period of reform and opening, Zhao emerged as an important leader at the top of the new administrative team. His policies in Sichuan became models for reform across the country. He was made premier of the State Council in 1980, a position long held by Zhou Enlai until his death in 1976. Zhao remained one of the three top leaders through the 1980s, and was a strong advocate of all-out marketization of the economy. In 1987 he became general secretary of the CPC, when Hu Yaobang was removed from that position. The debates over the extent to which markets should be allowed to drive the economy were intense in the 1980s. Hu and Zhao advocated the kind of "shock therapy" that would later prove so devastating to the post-Soviet economy in Russia.

Contradictions in the economy and within society arose as a result of the reforms, and in 1989 protests against the government led to the occupation of Tiananmen Square in central Beijing. Zhao supported the more radical demands of some protesters. As the movement became an anti-government effort and was manipulated by American and other foreign forces, Zhao resisted efforts to bring it to an end. After the PLA was brought in to restore order and normal activities in the capital, Zhao was removed from his offices. He spent his remaining years in forced retirement, living in Beijing and occasionally appearing playing golf or in discreet social settings. He wrote a memoir that was published in Hong Kong. He died in Beijing at the age of 85.

Zhou Enlai (1898-1976)

Zhou Enlai was born to a gentry landowning family in Jiangsu Province. He came of age at the time of the anti-imperialist May Fourth Movement (1919), during which he participated in printing a radical newspaper on campus at Nankai University in Tianjin. He was arrested the next year while leading a protest. He went to France with a worker-student program, and was part of the founding group of the Communist Party of China in Paris in 1921. He spent time in the Soviet Union before returning to China, where he became a key leader in the Jiangxi Soviet. Zhou backed Mao's selection as the main leader at the Zunyi Conference during the Long March in 1935, and remained a close supporter of the chairman from then on. He often served as the chief negotiator for the CPC in dealing with the Nationalists during and immediately after the War of Resistance Against Japanese Aggression (1937-45).

After the establishment of the People's Republic Zhou served as foreign minister and premier and was the architect of China's foreign policy for the next decades. Besides his natural talent at it, Zhou's close relationship with Mao made him particularly effective as a diplomat. Zhou negotiated successfully with India over Tibet. At the Geneva conference in 1954, he helped reach a settlement of the situation in Indochina, but was snubbed by US Secretary of State John Foster Dulles. In 1955, Zhou participated in the Afro-Asian conference of nonaligned countries at Bandung, in Indonesia. In 1971 Zhou met with Henry Kissinger to prepare for Nixon's historic visit to China which began the process of normalizing relations between the two countries. In January 1976 Zhou died after a long struggle with cancer. This left Chairman Mao without one of his most trusted and capable allies and advisors.

Zhu De (1886-1976)

Zhu De came from a poor peasant family in Sichuan, but was adopted by a wealthy relative at nine years of age. He was sent to study at the Yunnan Military Academy, setting him on the path of a military career. He was involved in the warlord affairs of China's southwest region until 1922 when he left the country to study in Germany. There he met Zhou Enlai and embraced Marxism and political activism. He was expelled from Germany for taking part in several protests, and returned to China where he joined the CPC.

Zhu became a leading figure in communist military activity. He collaborated with Mao Zedong in planning the 1927 Nanchang uprising, part of the efforts to revive party momentum following the split with the GMD in April of that year. When the uprising failed Zhu reconsolidated military units under his command, and in 1928 led his forces to join Mao at the base area in southern Jiangxi. This is considered the founding of the Red Army, with August 1 recognized as the anniversary of that event.

Throughout the 1930s and 1940s Zhu was the top leader of the Red Army and an important figure within the party leadership. He commanded the Eighth Route Army during the War of Resistance Against Japanese Aggression (1937-45) and the Civil War (1945-49). After liberation he was made commander-in-chief of the People's Liberation Army, as the Red Army was renamed, and in that capacity oversaw Chinese forces in Korea from 1950-53. In 1955 he was named one of the ten marshals of the PLA. Zhu died in July 1976.

ENDNOTES

1   Jonathan D. Spence, *The Search for Modern China* (New York: Norton, 1990), 122-23.

2   Angus Maddison, *The World Economy: A Millennial Perspective* (Paris: OECD Publications, 2001).

3   Yuan Xingpei, et al, eds., *The History of Chinese Civilization*, 4 vols. (Cambridge, U.K.: Cambridge University Press, 2012).

4   Yoshinobu Shiba. *Commerce and Society in Sung China*, trans. Mark Elvin, (Ann Arbor: Center for Chinese Studies, University of Michigan, 1992).

5   Ken Hammond. "Beyond the Sprouts of Capitalism: Toward an Understanding of China's Historical Political Economy and its Relationship to Contemporary China" *Monthly Review Online*, March 3, 2021, https://mronline.org/2021/03/03/beyond-the-sprouts-of-capitalism/; William Guanglin Liu. *The Chinese Market Economy, 1000-1500* (Albany: State University of New York Press, 2015).

6   Michael Marmé. *Suzhou: Where the Goods of All the Provinces Converge* (Stanford: Stanford University Press, 2005).

7   Jie Zhao, *Brush, Seal, and Abacus: Troubled Vitailty in Late Ming China's Economic Heartland, 1500-1644* (Hong Kong: Chinese University Press, 2018), 14; Timothy Brook, *The Confusions of Pleasure: Commerce and Culture in Ming China* (Berkeley: University of California Press, 1999); Craig Clunas, *Superfluous Things: Material Culture and Social Status in Early Modern China* (Honolulu: University of Hawaii Press, 2004).

8   Karl Marx, "Review of Guizot's Book on the English Revolution" in *Marx, Karl, Surveys from Exile*. (London: Verso, 2010), 254.

9   Margherita Zanasi, *Economic Thought in Modern China: Market and Consumption, c. 1500-1937* (Cambridge, U.K.: Cambridge University Press, 2020).

10 William T. Rowe, *Saving the World: Chen Hongmou and Elite Consciousness in Eighteenth-Century China* (Stanford: Stanford University Press, 2001); Helen Dunstan, *State or Merchant: Political Economy and Political Process in 1740s China* (Cambridge: Harvard University Asia Center, 2006).

11 The characterization of the political economy of the People's Republic of China in the present era, whether as "socialism with Chinese characteristics," "market socialism," "state capitalism," or in some other formulation, will be considered in chapter 5.

12 Immanuel C.Y. Hsü, *The Rise of Modern China*, 2nd ed. (Oxford, U.K.: Oxford University

Press, 1975), 169-188; Richard von Glahn, *The Economic History of China: From Antiquity to the Nineteenth Century* (Cambridge, U.K.: Cambridge University Press, 2016), 348-399.

13 William Dalrymple, *The Anarchy: The East India Company, Corporate Violence, and the Pillage of an Empire* (New York: Bloomsbury, 2019).

14 Carl A. Trocki, *Opium, Empire and the Global Political Economy: A Study of the Asian Opium Trade, 1750-1950* (London: Routledge, 1999).

15 Chang Hsin-pao, *Commissioner Lin and the Opium War* (Cambridge, MA: Harvard University Press, 1964); Joyce A. Madaney, *The Troublesome Legacy of Commissioner Lin: The Opium Trade and Opium Suppression in Fujian Province, 1820s to 1920s* (Cambridge, MA: Harvard University Asia Center, 2003).

16 John King Fairbank, *Trade and Diplomacy on the China Coast: The Opening of the Treaty Ports, 1842-1854* (Cambridge, MA: Harvard University Press, 1953).

17 Jean Chesneaux, Marianne Bastide, and Marie-Claire Bergère, *China: From the Opium Wars to the 1911 Revolution* (United States: Pantheon Books, 1977).

18 Jonathan D. Spence, *God's Chinese Son: The Taiping Heavenly Kingdom of Hong Xiuquan* (New York: Norton, 1996).

19 Louise Tythacott, ed., *Collecting and Displaying China's "Summer Palace" in the West: The Yuanmingyuan in Britain and France* (London: Routledge, 2019).

20 Lewis M. Chere, *The Diplomacy of the Sino-French War (1883-1885)* (Indiana: Crossroads, 1988).

21 Conrad Totman, *Early Modern Japan* (Berkeley: University of California Press, 1995).

22 William Beasley, *The Meiji Restoration* (Stanford: Stanford University Press, 2018).

23 Hsu, *The Rise of Modern China*, 406-428.

24 Luke S.K. Kwong, *A Mosaic of the Hundred Days: Personalities, Politics, and Ideas of 1898* (Cambridge, MA: Harvard University Press, 1984); Rebecca Karl and Peter Zarrow, eds., *Rethinking the 1898 Reform Period: Political and Cultural Change in Late Qing China* (Cambridge, MA: Harvard University Asia Center, 2002).

25 Joseph Esherick, *The Origins of the Boxer Uprising* (Berkeley: University of California Press, 1987).

26 Marie-Claire Bergère, *Sun Yat-sen* (Stanford: Stanford University Press, 2000).

27 Min Tu-ki, *National Polity and Local Power: The Transformation of Late Imperial China* (Cambridge, MA: Harvard University Press, 1989); Douglas R. Reynolds, *China 1895-1912: State-Sponsored Reforms and China's Late-Qing Revolution* (London: Routledge, 1996).

28 Chesneaux, et al, *China: From the Opium Wars to the 1911 Revolution*; G. Zay Wood, *The 21 Demands: Japan versus China* (Australia: Wentworth, 2019).

29 Lee Feigon, *Chen Duxiu: Founder of the Chinese Communist Party* (Princeton: Princeton University Press, 2014).

30 Wang Hui, *China's Twentieth Century: Revolution, Retreat, and the Road to Equality* (London: Verso, 2016), 8-109.

31 Elizabeth Forster, *1919, The Year that Changed China: A New History of the New Culture Movement* (Berlin: De Gruyter, 2019).

32 Margaret MacMillan, *Paris 1919: Six Months that Changed the World* (New York: Random House, 2003).

33 Chow Tse-tsung, *The May Fourth Movement: Intellectual Revolution in Modern China* (Stanford: Stanford University Press, 1967).

34 Lucien Bianco, *Origins of the Chinese Revolution, 1915-1949* (Stanford: Stanford University Press, 1971).

35 Jonathan D. Spence, *The Gate of Heavenly Peace: The Chinese and their Revolution* (New York: Penguin, 1982).

36 The history of the Communist Party can be traced through the documents and reports produced throughout the years of revolutionary struggle as gathered in Tony Saich, ed. *The Rise to Power of the Chinese Communist Party: Documents and Analysis* (Armonk, NY: M.E. Sharpe, 1996). Citations that follow indicate the relevant sections of this huge compendium.

37 For a clear-eyed analysis of the relative class forces in China in the early 1920s see Karl Radek's report of 22 June 1926, "On the fundamentals of Communist policy in China" in Alexander V. Pantsov, ed., *Karl Radek on China: Documents from the Former Secret Soviet Archives* (Chicago: Haymarket, 2021), 16-21.

38 Saich, *The Rise to Power of the Chinese Communist Party*, 3-100.

39 Lee Chae-jin, *Zhou En-lai: The Early Years* (Stanford: Stanford University Press, 1994); Alexander V. Pantsov and Steven Levine, *Deng Xiaoping: A Revolutionary Life* (Oxford, U.K.: Oxford University Press, 2015).

40 Saich, *The Rise to Power of the Chinese Communist Party*, 101-276.

41 S. A. Smith, *A Road is Made: Communism in Shanghai, 1920-1927* (London: Routledge, 2000).

42 For a fictionalized but sympathetic treatment of the 1927 GMD coup in Shanghai see André Malraux, *Man's Fate* (London: Penguin, 2009).

43 Saich, *The Rise to Power of the Chinese Communist Party*, 277-508.

44 Mao Zedong's writings are critical for understanding the development of the Chinese Revolution and of course for documenting his emergence as the key leader of the CPC in the 1930s and until his death in 1976. The most comprehensive edition of his writings from the 1920s through the establishment of the People's Republic of China in 1949 in English translation is Stuart R Schram, ed. *Mao's Road to Power: Revolutionary Writings 1912-1949* 5 vols, (Armonk, NY: M.E. Sharpe, 1992-1999). The volumes published by M.E. Sharpe took the project through 1937. Three subsequent volumes have been pub-

lished by Routledge from 1999-2015, bringing things up to 1945. Mao's "Report on the Peasant Movement in Hunan" delivered in February 1927 appears in Vol. II, 429-464. Citations to Mao's writings below will be from this series unless otherwise noted.

45 Saich, *The Rise to Power of the Chinese Communist Party*, 509-644.

46 Benjamin I. Schwartz, *Chinese Communism and the Rise of Mao* (Cambridge, MA: Harvard University Press, 1979).

47 Mao Zedong, *Report from Xunwu*, translated with an introduction by Roger R. Thompson (Stanford: Stanford University Press, 1990); Schram, *Mao's Road to Power*, vol. 3, 296-418.

48 Harrison Salisbury, *The Long March* (New York: Harper Collins, 1985).

49 In 1936 Mao Zedong held a series of interviews in Bao'an, Shaanxi, the temporary headquarters of the CPC, with the American journalist Edgar Snow in which he provided both an overview of the history of the Chinese revolution up to that point and an account of his own life and political activities. This remains one of the key sources on the period of the Jiangxi Soviet and the Long March. Edgar Snow, *Red Star Over China* (New York: Grove Press, 1973).

50 Mao Tse-tung (Zedong), *On Practice and On Contradiction*, intro by Slavoj Žižek London: Verso, 2017).

51 Chalmers A. Johnson, *Peasant Nationalism and Communist Power: The Emergence of Revolutionary China, 1937-1945* (Stanford: Stanford University Press, 1962).

52 Schram, *Mao's Road to Power*, vol. 8, 102-132.

53 Peter Duus et al, eds, *The Japanese Wartime Empire, 1931-1945* (Princeton: Princeton University Press, 1996).

54 Jean Chesneaux, Francoise Le Barbier, and Marie-Claire Bérgere, *China: From the 1911 Revolution to Liberation* (New York: Random House, 1977), 251-286.

55 Saich, *The Rise to Power of the Chinese Communist Party*, 853-970.

56 Gar Alperovitz, *The Decision to Use the Atomic Bomb* (New York: Vintage, 1996).

57 Saich, *The Rise to Power of the Chinese Communist Party*, 1185-1382.

58 Allan Shackleton, *Formosa Calling: An Eyewitness Account of the February 28th, 1947 Incident* (Manchester, U.K.: Camphor Press, 2017).

59 Denis Nowell Pritt, *Land Reform in China: How 500,000,000 Chinese Farmers, One Fifth of the World's People, Won the Right to their Own Land* (N.P.: Literary Licensing, 2012).

60 William Hinton, *Fanshen: A Documentary of Revolution in a Chinese Village* (New York: Vintage, 1966).

61 Lu Xun, *Jottings Under Lamplight* (Cambridge, MA: Harvard University Press, 2017), 256-261.

62 Ono Kazuko, *Chinese Women in a Century of Revolution* (Stanford: Stanford University Press, 1989), 176-188.

Endnotes

63 Elizabeth Mcguire, *Red at Heart: How Chinese Communists Fell in Love with the Russian Revolution* (Oxford, U.K.: Oxford University Press, 2018), 257-315.

64 Franz Schurmann, *Ideology and Organization in Communist China* (Berkeley: University of California Press, 1969).

65 Jean Chesneaux, *China: The People's Republic, 1949-1976* (New York: Pantheon, 1979), 31-82.

66 The development of the Struggle Between Two Lines was a highly complex process, followed most clearly in a book series by Roderick MacFarquhar. He was not sympathetic to the Chinese Revolution or the CPC, but his scholarly studies provide the best documentation of events available in English. Roderick MacFarquhar, *The Origins of the Cultural Revolution, Volume 1: Contradictions Among the People, 1956-57* (New York: Columbia University Press, 1974); Roderick MacFarquhar, *The Origins of the Cultural Revolution, Volume 2: The Great Leap Forward, 1958-59* (New York: Columbia University Press, 1983); Roderick MacFarquhar, *The Origins of the Cultural Revolution, Volume 3: The Coming of the Cataclysm, 1961-66* (New York: Columbia University Press, 1999).

67 Chesneaux, *China: The People's Republic*, 83-112.

68 Jurgen Domes, *Peng Te-huai: The Man and the Image* (Stanford: Stanford University Press, 1985).

69 Ellis Joffe, *Between Two Plenums: China's Intraleadership Conflict, 1959-1962* (Ann Arbor: University of Michigan Center for Chinese Studies, 1975).

70 These three texts were published in translation as Mao Tsetung (Zedong), *A Critique of Soviet Economics* (New York: Monthly Review, 1977).

71 The Anti-Rightist Campaign developed in response to some of the attacks on the CPC which emerged in the Hundred Flowers movement in 1956. This was an effort, initiated by Mao, to solicit appraisals and criticisms of the party's work in the first years after liberation. There were many initially positive responses, raising important issues but basically supporting the efforts going on to build a new socialist China. But then harsher attacks began to be launched calling for overthrowing the leadership of the party and essentially returning to the old social order. A number of people, many of them academics or other intellectuals, were criticized and some were sent to rural areas to take part in reeducation programs. It was some of these who were then used as convenient targets in 1962 to divert criticism from the abuses of power being indulged in by some local cadres.

72 Alessandro Russo, *Cultural Revolution and Revolutionary Culture* (Durham, NC: Duke University Press, 2020).

73 Jean Daubier, *A History of the Chinese Cultural Revolution* (New York: Vintage, 1974).

74 Jean Esmein, *The Chinese Cultural Revolution* (New York: Anchor, 1973).

75 Joan Robinson, *The Cultural Revolution in China* (London: Penguin, 1969).

76 Lee Hong Yung, *The Politics of the Chinese Cultural Revolution* (Berkeley: University of California Press, 1978).

77 C.F. Mobo Gao, *Gao Village: Rural Life in Modern China* (Honolulu: University of Hawaii Press, 2007); Dongping Han, *The Unknown Cultural Revolution: Life and Change in a Chinese Village* (New York: Monthly Review, 2008).

78 Edgar Snow, *The Long Revolution* (New York: Vintage, 1973), 179-190.

79 Ezra Vogel, *Deng Xiaoping and the Transformation of China* (Cambridge, MA: Harvard University Press, 2013).

80 Wen Tiejun, *Ten Crises: The Political Economy of China's Development* (1949-2020) (Singapore: Palgrave-Macmillan, 2021), 199-290.

81 "Review of the History of the Twenty-Eight Years Before the Founding of the People's Republic," Resolution on Certain Questions in the History of Our Party since the Founding of the People's Republic of China, adopted by the Sixth Plenary Session of the Eleventh Central Committee of the Communist Party of China, June 27, 1981, https://digitalarchive.wilson-center.org/document/resolution-certain-questions-history-our-party-founding-peoples-republic-china/.

82 Ching Kwan Lee, *Against the Law: Labor Protests in China's Rustbelt and Sunbelt* (Berkeley: University of California Press, 2007).

83 Cheng Li, ed., *China's Emerging Middle Class: Beyond Economic Transformation* (Washington, D.C.: Brookings, 2010).

84 Isabella Weber, *How China Escaped Shock Therapy: The Market Reform Debate* (London: Routledge, 2021).

85 Perhaps the best overview of the events of April-June 1989 is a documentary film by Carma Hinton and Richard Gordon, *The Gate of Heavenly Peace* (1995). Hinton was born in Beijing in 1949 and grew up in China until moving to the U.S. in the late 1970s. She made several documentary films about the village of Long Bow, Shanxi, which was the setting for her father's famous study of land reform, William Hinton, *Fanshen: A Documentary of Revolution in a Chinese Village* (New York: Vintage, 1966).

86 Bruce Gilley, *Tiger on the Brink: Jiang Zemin and China's New Elite* (Berkeley: University of California Press, 1998).

87 John Ross, *China's Great Road: Lessons for Marxist Theory and Socialist Practices* (Glasgow: Praxis, 2012), 225-234.

88 Hillary Clinton, "America's Pacific Century," *Foreign Policy* (October 2011).

89 As the General Secretary of the CPC and President of the People's Republic of China Xi Jinping has been the most important leader in China since his election in 2012. The best overall expression of his ideas about economic, political, and social issues are the three volumes (so far) of *The Governance of China* (Beijing: Foreign Language Press, 2014, 2017, 2020).

90 Barbara Finamore, *Will China Save the Planet?* (Cambridge, U.K.: Polity, 2018).

91 Deborah Brautigam, *The Dragon's Gift: The Real Story of China in Africa* (Oxford, U.K.: Oxford University Press, 2009).

92 Wang Hui, *The Politics of Imagining Asia* (Cambridge: Harvard University Press, 2011).

93 There have been many articles and commentaries on Xinjiang seeking to present a clearer picture of the situation there than that alleged by American and other Western politicians. Among the best of these is a report, Xinjiang: Understanding Complexity, Building Peace, *Instituto Diplomatico Internazionle, May 2021*, http://www.cese-m.eu/cesem/wp-content/uploads/2021/05/EN-Xinjiang-2.pdf (in English).

94 For a good overview of the 2019-20 events in Hong Kong, see "Hong Kong: From Royal Colony to Color Revolution," CodePink, September 8, 2021, https://www.codepink.org/hkweinar/.

95 K.J. Noh, "Preparing for War in the South China Sea," *LA Progressive*, July 18, 2016, https://www.laprogressive.com/war-and-peace/south-china-sea-war/.

96 The clearest indication of China's successful approach to managing the COVID-19 pandemic, putting people first, is in the basic statistics. With a population of 1.4 billion China, by May 2022, had 5,222 deaths from COVID-19. The US, with a population one-fourth the size of China's, had, by May 2022, more than 1,000,000 deaths, COVID-19 Data Repository by the Center for Systems Science and Engineering, Johns Hopkins University, https://github.com/CSSEGISandData/COVID-19/.

97 Yeo Yukyung, *Varieties of State Regulation: How China Regulates Its Socialist Market Economy* (Cambridge: Harvard University Asia Center, 2020).

98 Arthur G. Ashbrook, Jr., *An Economic Profile of Mainland China* (New York: Praeger, 1968), 18, quoted in E.L. Wheelwright and Bruce McFarlane, *The Chinese Road to Socialism* (New York: Monthly Review Press, 1970), 31.

99 Ji Chaozhu, *The Man on Mao's Right* (New York: Random House, 2008), 50.

100 Joan Robinson, *Reports from China* (London: Anglo-Chinese Educational Institute, 1977), 30.

101 Dongping Han, "The Socialist Legacy Underlies the Rise of Today's China in the World," *Aspects of India's Economy*, October 2014, https://www.rupe-india.org/59/han.html/.

102 Wilfred Burchett, *China's Feet Unbound* (London: Lawrence and Wishart, 1952), 40.

103 Burchett, *China's Feet Unbound*, 141.

104 Ibid., 142.

105 Wheelwright and McFarlane, *The Chinese Road to Socialism*, 35-6, 39.

106 Han Suyin, *Wind in the Tower: Mao Tsetung and the Chinese Revolution 1949-1975* (London: Jonathan Cape, 1976), 69.

107 Karl Marx, *The Civil War in France*, (Beijing: Foreign Languages Press, 1977), 66.

108 Han, *Wind in the Tower*, 124.

109 Historically, the Great Leap Forward is considered to be the 1958-1960 period, not the entirety of the five-year plan, because the CPC changed course prior to 1963.

110 Mao Tse-tung (Zedong), "Red and Expert," *Selected Works of Mao Tse-tung*, January 31, 1958, https://www.marxists.org/reference/archive/mao/selected-works/volume-8/mswv8_04.htm/.

111 V.I. Lenin, "Our Foreign and Domestic Position and Party Tasks," speech delivered to the Moscow Gubernia Conference of the R.C.P., November 21, 1920, *Collected Works*, 4th ed. (Moscow: Progress Publishers, 1966), 31:408-426.

112 Ibid., 419.

113 Han, *Wind in the Tower*, 129-130.

114 Wheelwright and McFarlane, *The Chinese Road to Socialism*, 40, 48.

115 Han, "The Socialist Legacy Underlies the Rise of Today's China in the World."

116 Han, *The Wind in the Tower*, 126-137; Dongping Han, *The Unknown Cultural Revolution* (New York: Monthly Review, 2008), 24-25; Jan Deleyne, *The Chinese Economy* (New York: Harper & Row, 1973), 21-22.

117 Han, *The Unknown Cultural Revolution*, 15.

118 Utsa Patnaik, "Revisiting Alleged 30 Million Famine Deaths during China's Great Leap," *MR Online*, June 26, 2011, https://mronline.org/2011/06/26/revisiting-alleged-30-million-famine-deaths-during-chinas-great-leap/.

119 Ibid.

120 Han, "The Socialist Legacy Underlies the Rise of Today's China in the World."

121 Sun Jingxian, "Population Change during China's 'Three Years of Hardship' (1959-1961)," *Contemporary Chinese Political Economy and Strategic Relations: An International Journal*, April 2016.

122 Han, *The Wind in the Tower*, 137.

123 Hongqi Editorial Department, *Long Live Leninism!* (Beijing: Foreign Languages Press, 1960), 1-10.

124 Wheelwright and McFarlane, *The Chinese Road to Socialism*, 69-71.

125 Hao Qi, "Distribution and Social Transition at Tonggang: China's Workers under Socialism, under 'Reform', and Today," *Aspects of India's Economy*, October 2014; Han, "The Socialist Legacy Underlies the Rise of Today's China in the World"; Stephen Andors, *China's Industrial Revolution: Politics, Planning, and Management, 1949 to the Present* (New York: Pantheon Books, 1977), 129; Han, *Wind in the Tower*, 179-180.

126 Qi, "Distribution and Social Transition at Tonggang"; Han, "The Socialist Legacy Underlies the Rise of Today's China in the World."

127 Han, "The Socialist Legacy Underlies the Rise of Today's China in the World."

128 The play was originally published in 1959.

129 Han, *The Wind in the Tower*, 263.

Endnotes

130 Han, *The Unknown Cultural Revolution*, 50; Han, *The Wind in the Tower*, 281.

131 Lin was the leader of the People's Liberation Army and named Mao's constitutionally designated successor during the Cultural Revolution.

132 Han, *The Wind in the Tower*, 284.

133 Mao Tse-Tung (Zedong), "A Letter to the Red Guards of Tsinghua University Middle School," *Long Live Mao Tse-tung Thought*, August 1, 1966, https://www.marxists.org/reference/archive/mao/selected-works/volume-9/mswv9_60.htm/.

134 "Decision Concerning the Great Proletarian Cultural Revolution," adopted on August 8, 1966, by the Central Committee of the CPC. Official English version published in R. Rojas, *La Guardia Roja Conquista China* (Chile: Editorial Prensa Latinoamericana, 1968), 430-40.

135 Mao Zedong, "Talk at a Meeting of the Central Cultural Revolution Group," *Long Live Mao Tse-tung Thought*, a Red Guard Publication, January 9, 1967.

136 Hongsheng Jiang, "The Paris Commune in Shanghai: The Masses, the State, and Dynamics of 'Continuous Revolution.'" Dissertation, Duke University, 2010, 408, https://dukespace.lib.duke.edu/dspace/bitstream/handle/10161/2356/D_Jiang_Hongsheng_a_201005.pdf/.

137 Ibid., 496-498.

138 Han, *The Unknown Cultural Revolution*, 60.

139 Ibid., 62-63.

140 American Friends Service Committee, *Experiment Without Precedent: Some Quaker Observations on China Today* (Philadelphia: AFSC, 1972), 20-21.

141 Janet Goldwasser and Stuart Dowty, *Chinese Factories Are Exciting Places!* (New York: Far East Reporter, 1973), 8, http://www.bannedthought.net/Magazines/FER/1973/FER-1973-ChineseFactoriesAreExcitingPlaces.pdf .

142 Ibid.

143 Joan Hinton, "How Can Socialism Ensure the Full Liberation of Women," Lecture delivered at Academy of Agricultural Mechanization Science, Beijing, 1997.

144 Ibid.

145 Goldwasser and Dowty, *Chinese Factories Are Exciting Places!*, 4; Han, *The Unknown Cultural Revolution*, 68-69; Joan Robinson, *Economic Management in China* (London: Anglo-Chinese Educational Institute, 1972), 16.

146 Han, *The Unknown Cultural Revolution*, 93-95.

147 Ibid., 95.

148 Maria Antonietta Macciocchi, *Daily Life in Revolutionary China* (New York: Monthly Review Press, 1972), 250.

149 John O. Killens, *Black Man in New China* (Los Angeles: US-China Peoples Friendship Association, 1976).

150 George W. Crockett, Jr., "People's Justice: Judge George W. Crockett, Jr., looks at China's Legal System," *New China*, June 1976, 27.

151 Macciocchi, *Daily Life in Revolutionary China*, 275-282.

152 Neil Conner, "The Cave the Chinese President Called Home," *The Telegraph*, October, 19, 2015, https://s.telegraph.co.uk/graphics/projects/xi-jinping-cave/index.html; Han, *The Unknown Cultural Revolution*, 95; Macciocchi, *Daily Life in Revolutionary China*, 27.

153 Macciocchi, *Daily Life in Revolutionary China*, 54

154 Han, *The Unknown Cultural Revolution*, 69.

155 Mao Tse-tung (Zedong), *On Contradiction* (Beijing: Foreign Languages Press, 1967), 314.

156 Macciocchi, *Daily Life in Revolutionary China*, 152-153.

157 Han, *The Unknown Cultural Revolution*, 125.

158 Ibid., 122, 124.

159 Ibid., 122-123.

160 Macciocchi, *Daily Life in Revolutionary China*, 208, 212, 214.

161 William Hinton, *The Great Reversal: The Privatization of China • 1978-1989* (New York: Monthly Review, 1990), 144.

162 Mobo Gao, "Why Is the Battle for China's Past Relevant to Us Today?," *Aspects of India's Economy*, October 2014.

163 Ruth Sidel, *Women and Child Care in China* (Baltimore: Penguin, 1973), 23-24.

164 American Friends Service Committee, *Experiment Without Precedent*, 24-25.

165 Harry Belafonte and Helen Rosen, "Harry Belafonte: An Exception Wants to Change the Rule," *New China*, June 1976, 17, http://www.bannedthought.net/China/MaoEra/ContemporaryCommentary/US-China-PFA/NewChina/NewChina-V2N1-1976-June.pdf/.

166 Imari Obadele, *Free the Land!* (Washington DC: House of Songhay, 1984), 18.

167 Hinton, *The Great Reversal: The Privatization of China • 1978-1989*, 141.

168 Cheng Zhongyuan, *The Era of China's Turning Point, Deng Xiaoping in 1975-1982* (New York: Portico, 2018), part I.

169 Mobo, "Why Is the Battle for China's Past Relevant to Us Today?"

Index

A

ACFTU (All-China Federation of Trade Unions), 84, 130
Afghanistan, 87–88
Africa, 6–7, 21, 80–81, 149
agricultural collectivization, 41, 44–45, 47–48
agricultural sector, 4, 44, 63–64, 66, 97
agriculture, 2, 5, 68, 95–97, 99–100, 102, 149
Alassane Diop, 123
Alibaba, 91
All-China Women's Federation, 40, 130
alliance, 4, 14, 25, 30, 112, 140
alternative energy, 79, 81
Amiri Baraka, 124
Angola, 126
Anshan Constitution, 106
anti-imperialism, 93, 123
APCs (agricultural producers' cooperatives), 45
armies, people's, 29, 35, 145
Asia, 6, 21, 23, 80–81, 149

B

Basic Law, 88–89
Beijing, 13–14, 18–19, 22–23, 34–36, 41, 52–54, 56–58, 69–72, 108–9, 113, 130–31, 134–35, 137–40, 148, 150
Beijing University, 19, 23, 53, 108, 145–46
Belafonte, Harry, 123
Belt and Road Initiative (BRI), 74, 81, 149
Bengal, 7
big character posters, 53, 107–9, 113
Black Panthers, 124
Bo Gu, 30
Bolshevik Revolution, 18–19, 22, 132, 136, 139
Bolsheviks, 23–25, 111
Boxer Rebellion, 13–14, 131, 134
Britain, 1, 7–8, 11, 13, 21, 88, 140

C

campaigns, 16–17, 29, 44, 90, 97, 121, 145
political, 44, 83, 121
cancer, 33, 136, 143, 151
capital, 1, 10–11, 22, 32, 34, 47, 72, 75, 88, 137, 139
private, 91, 95
capitalism, 2–3, 12, 23, 69, 98, 106, 111, 125

capitalist road, 110, 131, 133, 143
capitalist roaders, 107–8, 110, 113
capitalists, 4, 21, 76, 83, 98, 125, 141
Central Committee, 100, 109–11, 142
Central Cultural Revolution Group, 131, 135
central government, 90, 104, 137
Central Party School, 148
chairman, 45, 50, 78, 106–7, 134, 139, 142–43, 150
Chen Boda, 131
Chen Duxiu, 18, 23–24
Chen Yonggui, 126
Chiang Kai-shek (Jiang Jieshi), 26–27, 32–34, 132, 140–41
China, new, 18–19, 34, 36–37, 43, 48, 123, 130, 146
China's development, 40, 68, 88
China's economy, 3, 44, 75
China's Revolution, 2–150
China's Socialist Revolution, 95–127
Chinese Communist Party, 24, 93, 95, 120
Chinese Nationalist Party, 14
Chinese People's Volunteers, 42, 145
Chongqing, 33, 133
cities, 2–3, 18–19, 26–28, 33–34, 49, 51–53, 55, 58, 64–65, 68–69, 71–72, 117, 119, 144, 146–48
civil war, 23, 34–35, 37, 55, 94, 132–33, 138, 141, 145, 151
collectivization, 44–45, 101
Commune, Paris, 54, 105, 112, 148
communes, 46, 49–50, 54–55, 101–3, 109, 115, 119, 123, 126–27, 135
communism, 62, 99–100, 103, 111, 123
Communist International, 23–24, 41, 136, 139, 147
Communist Party (CPSU), 23–25, 37, 41, 58, 62, 69–70, 72, 78, 130, 132–34, 136–40, 148, 150
Communist Youth League (CYL), 138, 141, 148–49
Confucius, 121, 129
contradictions, 1, 5–6, 10, 44, 57–58, 63, 67, 72–73, 75, 84, 86, 91, 93
corruption, 43–44, 67, 69, 77, 79, 81
COVID-19 virus, 91
CPC (Communist Party of China), 24–25, 27–29, 31–32, 34, 36–37, 40–41, 47–48,

50–52, 54–55, 58, 61–62, 78–79, 90–91, 93, 130–31, 133–36, 138–44, 148–51
CPPCC (Chinese People's Political Consultative Conference), 43, 132–33
Cuba, 111
Cultural Revolution, 51–55, 58, 107–8, 111, 113–15, 117–19, 121–22, 124–25, 130–31, 135, 142, 145–46, 148–49

D
Dalai Lama, 86
defense, minister of, 50, 142, 145
Deng Xiaoping, 24, 47, 59–75, 107, 127, 133–34, 136, 138, 143, 149
doctors, barefoot, 55, 119, 123
Dongping Han, 103, 107, 113, 120
DPRK (Democratic People's Republic of Korea), 42
dynasties, 10, 13–15, 85, 134

E
economic development, 25, 31, 43, 45, 62, 67, 81–82, 86–88, 105, 141
economic growth, 59, 66, 82, 97–98, 100, 115
economic reforms, 72, 76, 149
economy, 24, 61–62, 73, 75–77, 79, 82, 95, 97, 126–27, 130, 138, 140, 149–50
productive, 62, 77, 134, 140–41
urban, 43–44
education, 62, 64, 77, 82, 85–86, 101–2, 106, 108, 117, 126, 128, 130
EIC (East India Company), 6–7
Eight-Power Expeditionary Force, 13–14
elections, 31, 88–89
Empress Dowager Cixi, 13, 134
English Revolution, 4
enterprises, 43, 48, 65, 80, 91
export, 45, 73, 77

F
factories, 9, 11, 19, 22, 54, 96, 102, 106–7, 109, 112, 116, 119, 123
families, 28, 31, 51, 57, 64, 66, 68, 116, 132–33, 138, 143
famine, 94, 97, 103–4
farmers, 7, 24–25, 28, 45, 120, 143
food, 46, 49–50, 94, 106, 128

food crisis, 49–50
France, 11–12, 14, 21–22, 24, 32, 124, 133, 140, 144, 147, 150
fronts, united, 25–26, 99, 132, 136, 142, 147

G
Gang of four, 58, 135
Germany, 14, 21–22, 42, 151
global capitalist system, 12, 75–76, 141
GMD, 14, 23–27, 29, 32, 35, 41, 132, 136–37, 141–42, 147, 151
Gorbachev, 70
government, 16, 18, 22, 40, 42, 44, 61–63, 69–73, 75–76, 78–79, 84–87, 105–6, 130, 132–35, 140
Great Leap Forward, 45–46, 48, 51–52, 63, 100–107, 123, 125, 128, 135, 142–43, 145
Great Proletarian Cultural Revolution, 107, 110, 128, 143
Guangxu, 134
Guangxu Emperor, 12–13, 15, 129, 134
Guangzhou, 1, 7–9, 26, 65, 117, 137, 141

H
health, public, 79, 86, 91
health care, 55, 64, 77, 82, 101–2, 106, 118–19, 126, 128
home, 4, 17, 22, 39, 56, 83, 116, 123, 146
Hong Kong, 65, 88–89, 147, 150
Hong Xiuquan, 10, 137
housing, 62, 65, 77, 82, 89, 91, 115, 126
Huaihai battle, 35
Huaihai Campaign, 35
Hubei provinces, 26, 141
Hu Jintao, 78, 137, 141
Hunan, 2, 27, 138, 142–44
Hu Qiaomu, 69
Hu Yaobang, 68–70, 138, 150

I
ideology, 93, 107, 110, 147
imperial examinations, 5, 129, 137
imperial system, 2, 5, 14, 16, 18, 23, 38, 136
incomes, 52, 63, 67, 82, 94
India, 6–7, 32, 42, 86, 124, 150
International Concession, 26
isolation, 101, 123, 126

Index

J

Japan, 12–14, 16, 22, 32–34, 41, 57, 129–30, 132, 141, 144, 147
Japanese Aggression, 32, 37, 145, 150–51
Japanese forces, 33–34
Japanese occupation, 33
Jiang Jieshi, 26
Jiang Qing, 58, 61, 131, 135–36
Jiang Shangqing, 140
Jiangsu Province, 137, 140, 150
Jiangxi, 27, 132
Jiangxi base area, 29
Jiangxi Soviet, 28–29, 141, 143, 145, 150
Jiang Zemin, 72, 75, 137–38, 140
Ji Chaozhu, 94
Jinping, 74, 78–79, 81, 91, 138, 148

K

Kang Youwei, 12, 129, 147
Khrushchev, 49, 51, 99, 104
Killens, John, 124
Kissinger, Henry, 151
Korea, 12, 16, 32, 41–42, 44, 145, 151

L

labor, 4, 19, 28, 30, 45, 47, 97, 99, 101, 117, 130
land, 4, 12, 28, 38–40, 45, 62, 64, 68, 85, 93, 96
landlords, 19, 38–39, 51, 93–94, 96
land reform, 29, 31, 38–40, 49, 86, 93, 95
Latin America, 80–81, 149
leadership, 27, 29–30, 47, 49–50, 53, 55, 58, 70, 74, 133–35, 140, 143, 147–48
Lenin, 19, 23, 31, 92, 98, 100, 111
Liang Qichao, 12–13, 129–30
liberation, 21–37, 40–41, 89, 93, 130, 133, 135, 140, 142, 145, 148–49, 151
Li Dazhao, 23–24
Li Lisan, 24, 30
Lin Biao, 56, 109, 111, 125, 141, 145
Lin Zexu, 8
Li Peng, 69
Liu Shaoqi, 50, 107, 133, 142–43
Long March, 29–31, 107, 132–33, 138, 141–43, 145, 150
Lushan Plenum, 49, 52, 141–43, 145
Lu Xun, 18, 39

M

mainland, 35, 89–90, 95, 136, 138, 141
management, 43–44, 47, 50, 54, 65, 106, 145
Manchuria, 32–34
Mao, 27–28, 30–31, 41, 47, 50–56, 58, 63–64, 101–2, 107–9, 111, 120–21, 127, 135, 142–43, 145, 150–51
Mao era, 93, 96, 107, 122, 125, 127
Mao's leadership, 63, 142
Mao Zedong, 19, 23–25, 37–38, 57, 61, 95, 110, 123, 131, 133–34, 138, 142–43, 145
Maring, 24, 139
market mechanisms, 62, 67, 73, 90
markets, 1–4, 9, 62, 65, 134, 149–50
marriage, 40, 102
Marriage Law, 39–40, 96
Marx, 4, 23, 30–31, 92, 98, 100
Marx, Karl, 19
Marxism, 23, 105, 123, 144, 149
massacres, 72, 114, 132
masses, 9–10, 13, 28–29, 45, 48, 51–52, 54–55, 58, 100, 106–7, 109–11, 114, 148
mass mobilizations, 47, 102, 122
material standards, 52, 63–64, 66–67, 75, 80
May Fourth Movement, 22–23, 25–35, 143–44
military, 15, 31, 56, 114, 145, 147
Ming Dynasty, 3, 52, 108, 137
Mobo Gao, 122
Modern China, 114, 127
Moscow, 23, 41, 133, 139, 141–42
movement, 10, 18, 21, 72, 76, 108, 111, 131, 135, 137, 139, 146, 150
Mozambique, 124

N

Nanjing, 3, 8–10, 13, 33, 85, 137
Nationalist Party, 14, 23, 25
Nationalists, 16, 25, 29, 32–36, 88, 105, 147, 150
National People's Congress (NPC), 133, 142, 148
negotiations, 26, 34, 41, 90, 139, 141
New Cold War, 82–83
newspapers, 40, 52, 102, 112
Ninth Party Congress, 55–56, 58, 131, 135, 142
Nixon, 57, 151

O
opium, 7–8
Opium Wars, 8, 12, 88

P
Paris, 19, 21, 133, 144, 150
party, 25–32, 40, 47–50, 52–55, 58, 61–63, 69–70, 72–73, 76, 78–79, 91, 106–9, 113–14, 116–17, 135–38, 140, 142–43, 147–48
party leaders, 58, 70–71, 109, 111, 145
party leadership, 29, 40, 48, 72, 108, 138, 142, 151
party members, 48, 76
party secretary, 53, 133, 137–38, 140, 149
party work, 138
Peasant associations, 101
peasants, 18–19, 27–28, 38–39, 41, 51, 93–98, 101–2, 104, 108, 111–12, 114–15, 119, 121–22
Peng Dehuai, 50, 138, 141, 144
People's Communes, 46, 96
People's Republic, 36, 38, 57–58, 69, 72, 75, 77, 84–86, 88, 90–91, 96, 130, 132
People's Republic of China. See PRC
Phases of China's Socialist Revolution, 95–127
PLA (People's Liberation Army), 35, 40, 47–48, 55–56, 71–72, 76, 109, 114, 116, 119, 139, 141–42, 145, 148, 150–51
plenum, 50, 61, 111
policies, 47, 52, 58, 63, 66–68, 72–73, 78, 86, 88, 133, 138, 140, 144
Political Bureau, 30, 69, 138–39
political economy, 3, 51, 120
population, 4, 17, 24, 37, 61, 63–64, 66, 84, 94–95
ports, 8, 12, 120, 124
posters, 46, 53, 70, 111, 146
power, 10, 12–13, 19, 21–22, 26–27, 34, 36, 38–39, 44, 67–69, 84–86, 88–90, 112–14, 133, 146
foreign, 13, 26, 93
political, 3, 84, 112, 114
PRC (People's Republic of China), 22, 24, 37, 41–42, 48, 50, 61, 63, 66, 70, 73, 75, 78–79, 132–33, 142–44
privatization, 66, 68

production, 2, 4, 43, 45, 47, 62, 64, 77, 79, 97, 99, 102, 104, 106–7, 120
productivity, 5, 44–45, 48, 62, 98, 107, 120, 125
profits, 5, 73, 84, 91, 125
Proletarian Cultural Revolution, 52, 107, 110
proletariat, 24–25, 110

Q
Qianlong emperor, 1–3, 6–7
Qing, 10, 12–13, 18, 136–37, 146–47
Qing dynasty, 1, 3, 10–11, 15, 32, 85–86, 129, 136–37, 147
Qing government, 8, 10, 13–14, 16, 131
Qinghua University, 69, 137, 148

R
rebellions, 5, 9, 11, 27, 35, 86, 111, 137
Red Army, 27, 29, 31, 33–35, 40–41, 132–33, 138, 143, 145, 149, 151
Red Guards, 53–55, 109, 114, 145–46, 148
reform era, 61, 63, 70, 73, 75
reform movement, 13, 129
reform policies, 66, 72–73
reforms, 12–13, 15, 40, 44, 61–77, 79, 82, 129, 134, 138, 140, 144, 149–50
Republic of China government, 132
revolution, 27–28, 30–32, 34–35, 63, 93–95, 97, 105, 114, 120, 122, 125, 127, 139, 146–47, 149
revolutionary base areas, 29, 31
Revolutionary Committees, 55, 113–16, 130, 135, 147–48
Robeson, Paul, 118, 123, 124
Robinson, Joan, 94, 124
Russia, 14, 18, 23, 32, 136, 150

S
Second Opium War, 10, 134
Second United Front, 32–33, 132
Self-Strengthening Movement, 11, 13, 146–47
SEZs (special economic zones), 63, 65–66
Shandong, 22, 32, 131
Shandong province, 10, 13, 45, 113, 131
Shanghai, 9–10, 13, 23–24, 26–27, 32–33, 52, 54, 112, 129, 132, 135–36, 140, 142, 148
Shanghai Commune, 54, 112, 131, 135, 148

Index

Shanghai Communique, 57
Shirley Graham DuBois, 124
Sichuan, 2, 31, 64, 149, 151
Sichuan province, 33, 133, 149
Sino-Soviet Friendship Treaty, 40
Sino-Soviet Split, 41, 143
slogan, 40, 97, 121, 130
socialism, 24, 62–63, 91, 93, 98–100, 102, 106, 120, 125, 144, 147–48
socialist camp, 56, 98–99, 125
socialist construction, 48, 62, 98, 100, 107, 125
socialist development, 41, 45, 51, 134
socialist planning, 95, 98
socialist road, 98, 105, 107, 109
socialist system, 110, 127
social life, 46, 52, 72, 102
SOEs (state-owned enterprises), 63–64, 66–68
soldiers, 15, 71
Songtao Reservoir, 101
South China Sea, 88, 90
Soviet Communist Party, 41, 49
soviets, 41, 48–50, 97, 99–100, 105, 126, 141, 143
Stalin, 28, 41, 51
students, 22, 24, 53–54, 69–70, 108–11, 113, 119, 121, 141, 144–45
Sun Yatsen, 14–16, 24, 26, 132, 136, 147

T

Taiping movement, 9
Taiping Rebellion, 9–10, 134
Taipings, 10, 137
Taiping Tianguo, 10, 137
Taiwan, 22, 32, 35, 37, 43, 57, 66, 88–90, 132, 138
Tan Sitong, 12–13, 129
teachers, 55, 67, 110–11, 117, 124
technology, 73, 80, 102, 133
Third International, 23
Tiananmen Square, 22, 53, 58, 69–71, 146
occupation of, 72, 150
Tibet, 84–87, 150
Tibetan society, 86
Tibet Autonomous Region, 138
Tongmenghui, 14, 136
Tongzhi Restoration, 134
towns, 5, 29, 34, 44, 100
TPP (Trans-Pacific Partnership), 82

Treaty of Nanjing, 8–9, 13
Trump, Donald, 82
Twenty-One Demands, 16, 32
Two-Line Struggle, 105

U

Uighurs, 84–85, 87
unions, 27, 84, 97
United Kingdom, 124
United Nations, 42, 82, 88, 133
United States, 21, 33–34, 56–57, 75, 78, 86, 88–90, 94, 99, 139–42, 149
universities, 67, 69, 104, 119
uprisings, 26–28, 151
US, 14–15, 33, 42–43, 56, 70, 73, 76–77, 79–80, 83, 88–90, 111
USSR (Union of Soviet Socialist Republics), 49, 51, 97–98, 105–6, 124, 126, 145
Ussuri River, 56
Uzbeks, 84, 87

V

Versailles, 19, 21–22, 144
Vietnam, 56, 90
Vietnam war, 124
villagers, 96, 121
villages, 38, 40, 45–46, 51–52, 55, 64, 66, 77, 96, 101, 112–13, 115–17, 146
rural, 26, 40, 119, 121

W

wages, 27, 34, 66, 105, 115
Wang Hongwen, 58, 61, 131, 135
war, 12, 16, 21–22, 24, 32–34, 41–44, 124, 132, 141, 144–45, 147
warlords, 17–19, 93
War of Resistance, 37, 42, 145, 150–51
Washington, 42, 89
water, 68, 79, 88, 90, 97, 101, 108
Wenhui Bao, 108
Whampoa Military Academy, 26–27, 139, 141
woman, 39, 96, 115, 117, 123
women, 18, 28, 39–40, 96, 100, 102, 115–16, 123, 130–31
workers, 3–4, 18–19, 26–28, 54, 65–67, 77, 83–84, 93, 98–99, 102, 104, 106–7, 110–12, 114–15, 117, 120, 125–26, 130

working class, 25, 51, 55, 72, 84, 100, 104–6, 109, 111–13
World Bank, 81
World Trade Organization, 75–76, 141
World War I, 21–22, 24, 32, 144
World War II, 14, 118, 140

X
Xi'an, 32, 132
Xinhua, 80
Xinjiang, 80, 84–87

Y
Yalu River, 42

Yan'an, 31–32, 34, 54, 132, 135, 138, 142–43
Yan'an Base Area, 29, 138
Yao Wenyuan, 52, 58, 61, 108, 131, 135
Yuan Shikai, 13, 16, 147

Z
Zambia, 124
Zhang Chunqiao, 58, 61, 131, 135
Zhao Ziyang, 70, 72, 139–40, 149
Zhejiang province, 24, 132
Zhengzhou, 115
Zhou Enlai, 30, 32, 57–58, 133, 141, 149–51
Zimbabwe African National Union, 124
Zunyi Conference, 29, 31, 133, 142–43, 150

Printed in the USA
CPSIA information can be obtained
at www.ICGtesting.com
CBHW020407140624
9995CB00004B/22